Advanced Technologies for Industrial Applications

Rohit Thanki • Purva Joshi

Advanced Technologies for Industrial Applications

Springer

Rohit Thanki
Krian Software GmbH
Wolfsburg, Germany

Purva Joshi
Department of Information Engineering
University of Pisa
Pisa, Italy

ISBN 978-3-031-33237-1 ISBN 978-3-031-33238-8 (eBook)
https://doi.org/10.1007/978-3-031-33238-8

This Springer imprint is published by the registered company Springer Nature Switzerland AG
The registered company address is: Gewerbestrasse 11, 6330 Cham, Switzerland

We are honored to dedicate this book to our family, gurus, and loved ones. You have been our unwavering source of support, inspiration, and blessing throughout our life.

Preface

Our modern society cannot ignore the influence of technology. We are immersed in technology if we are at a desktop or laptop in the office, checking heart rates from our smartwatches, playing with our iPhone, or talking to Alexa. Every industry is being disrupted by technology, and we all know that. Even relatively new areas such as the development of new tools and applications are being disrupted by the next generation of technology, which moves incredibly fast. Moore's law state that transistors will continue to double on integrated circuits every year in the near future. As technology advances, it becomes more powerful and faster while simultaneously becoming more lightweight, and it is happening at an alarming rate.

In this book, we discussed a variety of technologies such as system identification, signal processing, computer vision, and artificial intelligence and their usage in industries. These technologies have great market values and significant influence on human society. Various tools and applications have been developed using these technologies to better human culture. During the pandemic, technologies became critical assets for developing modern industrial applications. This book covers the usage and importance of these technologies in various industrial applications. Also, this book provides future technological tools which help in the development of a variety of industrial applications.

In Chap. 1, basic information of various technologies which are used in industrial applications. Chapter 2 addresses a basic concept of system identification and its usage in various industries. In Chap. 3, we present a signal processing and its applications in various areas such as broadcasting, defense, etc. Chapter 4 gives information regarding computer vision technology and its usage in various industries. Furthermore, artificial intelligence technology along with its commercial usage are provided in Chap. 5. Chapter 6 gives advanced technological tools based on technology such as Internet of Health Things, autonomous robots, etc. The book has following features:

- Describes basic terminologies of various technologies such as system identification, signal processing, computer vision, and artificial intelligence
- Presents various technological tools for industrial applications

- Gives usage of system identification and artificial intelligence in industrial applications
- Provides technical information on upcoming technologies for industrial applications

Our task has been easier and the final version of the book considerably better because of the help we have received. We would also like to thank Mary James, Executive Editor, Springer, and other representative of Springer, for their helpful guidance and encouragement during the creation of this book.

Wolfsburg, Germany
Pisa, Italy
March, 2023

Rohit Thanki
Purva Joshi

Contents

Chapter 1
Introduction

Technology has become an integral part of our daily lives, and it's hard to imagine a world without it. Technological advancements have changed how we communicate, work, and interact with the world around us, from smartphones to social media, from artificial intelligence to the Internet of Things. Technology has revolutionized nearly every industry, from healthcare to finance, education to transportation.

In this day and age, it's important to understand the impact of technology on our lives and society as a whole. With new technologies emerging daily, keeping up with the latest advancements and understanding how they work can be challenging. But by staying informed and aware of the benefits and challenges of technology, we can make informed decisions about its use and create a better future for ourselves and our communities. The technology discussed in this book will explore the latest advancements, their potential applications, and the ethical considerations surrounding their use. We'll look at how technology has transformed industries, from healthcare to finance, and we'll consider the ways in which it is likely to shape our world in the years to come. By the end of reading this book, you'll better understand the role technology plays in our lives and the importance of staying informed about the latest advancements.

Technological advancements have brought about significant changes in various industries, enabling them to operate more efficiently and effectively. The use of technology has become a crucial aspect of modern-day industries, as it helps to improve productivity, quality, and speed while reducing costs. From automation to artificial intelligence, machine learning to the Internet of Things, industries use the latest technologies to streamline operations and gain a competitive edge. By leveraging these tools, industries can optimize their supply chains, manage inventory more effectively, and monitor their production processes to ensure they run efficiently.

This book will explore how industries use technology to transform operations and achieve goals. We will look at the latest advancements in automation, robotics, and other technologies and consider their potential applications in various industries,

R. Thanki, P. Joshi, *Advanced Technologies for Industrial Applications*,
https://doi.org/10.1007/978-3-031-33238-8_1

such as manufacturing, healthcare, transportation, and more. We'll also examine the challenges and ethical considerations surrounding the use of technology in the industry and how businesses can navigate these issues to ensure responsible and sustainable growth. By reading this book, you'll better understand how technology transforms industries and how businesses can leverage these advancements to remain competitive and drive growth.

Technologies have brought about significant changes in the industry, and their impact cannot be overstated. Here are some of the key ways in which technologies are significant in the industry:

1. **Improved Efficiency:** By using technologies such as automation, machine learning, and robotics, industries can optimize their processes and reduce the time and resources required to complete tasks. This leads to increased productivity, faster production cycles, and lower costs.
2. **Better Quality:** Technologies such as sensors and analytics enable industries to monitor their production processes and identify areas for improvement. This leads to better quality products and services and improved safety and compliance.
3. **Increased Flexibility:** Technologies like 3D printing and additive manufacturing enable industries to produce complex parts and products more quickly and flexibly. This allows for more customization and faster time to market, which can be a significant competitive advantage.
4. **Enhanced Safety:** Technologies such as drones and remote monitoring enable industries to conduct inspections and maintenance activities in hazardous or hard-to-reach areas without putting workers at risk. This leads to increased safety for workers and reduced downtime for the business.
5. **Improved Sustainability:** Technologies such as renewable energy, energy storage, and recycling enable industries to reduce their environmental impact and operate more sustainably. This can be a significant differentiator for businesses that want to appeal to environmentally conscious consumers and stakeholders.

Overall, the significant impact of technologies in the industry is that they enable businesses to operate more efficiently, effectively, and sustainably, leading to increased profitability and growth. By leveraging the latest technologies, businesses can stay competitive and meet the market's ever-changing demands. Technologies are used extensively in various industries to streamline operations, increase productivity, and reduce costs. Here are some examples of how technologies are being used in industry:

- **Automation:** Automation technologies, such as robots and conveyor systems, are used in manufacturing to perform repetitive and dangerous tasks. This leads to increased efficiency and safety for workers.
- **Additive Manufacturing:** Additive manufacturing technologies, such as 3D printing, are used to produce complex parts and products more quickly and with greater flexibility. This enables industries to customize products and reduce time to market.

- **Machine Learning:** Machine learning technologies are used to analyze large amounts of data and identify patterns and insights. This enables industries to make better decisions, optimize processes, and improve quality.
- **Internet of Things:** The Internet of Things (IoT) connects machines and devices, enabling industries to monitor and manage their operations more effectively. This leads to improved efficiency, reduced downtime, and better quality.
- **Renewable Energy:** Renewable energy technologies, such as solar and wind power, are used in various industries to reduce reliance on fossil fuels and operate more sustainably.
- **Augmented Reality:** Augmented reality technologies are used in industries such as healthcare and education to enhance learning and training experiences.
- **Blockchain:** Blockchain technologies are being used in industries such as finance and supply chain management to increase transparency and security.

Overall, the usage of technologies in the industry is vast and varied, and new technologies are being developed and applied all the time. By adopting the latest technologies, industries can stay competitive and meet the market's changing demands.

This book contains six chapters that cover most emerging technologies used in real applications and different industries worldwide. Chapter 1 gives a broad view of different technologies and their significance in the industry. Chapters 2–6 give information on various technologies such as system identification, signal processing, image processing, artificial intelligence, and advanced technologies.

System identification is the process of building mathematical models of physical systems using measured input and output data. These models can be used to understand and predict the system's behavior and design control systems to achieve desired performance. System identification is a critical tool in fields such as engineering, physics, economics, and biology and is used in a wide range of applications, including control of aircraft, optimization of energy systems, and modeling of biological processes. In this way, system identification provides a powerful framework for understanding and controlling complex systems and has important implications for many areas of science and engineering. All information regarding system identification is covered in Chap. 2.

Signal processing is a broad field of study that involves the analysis, modification, and synthesis of signals, which are patterns of variation that convey information. Signals can take many forms, including audio, video, images, and other data types. Signal processing is critical in many fields, including communications, image and video processing, audio processing, and control systems. The goal of signal processing is to extract meaningful information from signals and to use that information to make decisions or take action. This involves techniques such as filtering, smoothing, and compression and more advanced methods such as machine learning and artificial intelligence. Signal processing has many applications, including speech and audio processing, medical imaging, radar and sonar, and telecommunications. It is essential to many modern technologies, such as smartphones, streaming media services, and autonomous vehicles. In summary, signal processing is a powerful and versatile field of study that plays a critical role in many areas of science and technology. By analyzing and manipulating signals, signal processing allows us to extract information and make decisions that can improve our lives and advance our understanding of the world around us. All information regarding signal processing is covered in Chap. 3.

Image processing is a field of study that involves the analysis and manipulation of digital images. Digital images are composed of pixels, each representing a single point of color or intensity within the image. Image processing techniques can be used to enhance or modify images, extract information from them, or perform other tasks such as compression and transmission. Image processing has many applications in fields such as medicine, remote sensing, and computer vision. In medical imaging, for example, image processing techniques can be used to enhance images of the human body for diagnostic purposes. In remote sensing, image processing can be used to analyze satellite imagery to monitor environmental changes or detect objects on the ground. In computer vision, image processing enables machines to "see" and interpret the visual world. Image processing techniques range from simple operations such as resizing and cropping to more advanced methods such as image segmentation, feature extraction, and machine learning. These techniques can be applied to images from various sources, including digital cameras, medical imaging equipment, and satellites. In summary, image processing is a powerful and versatile field of study that allows us to analyze, modify, and extract information from digital images. With applications in fields such as medicine, remote sensing, and computer vision, image processing is essential for advancing our understanding of the world around us. All information regarding image processing is covered in Chap. 4.

The field of artificial intelligence (AI) is one of the fastest-growing academic fields as it aims to create machines that can perform tasks that normally require human intelligence. This includes learning, problem-solving, decision-making, and language understanding tasks. AI can transform many industries, from healthcare to finance to transportation. AI aims to create machines that can learn and adapt to new situations as humans do. This involves developing algorithms and models to analyze large amounts of data and make predictions based on that data. Machine learning, a subfield of AI, is compelling, allowing machines to learn from experience and improve their performance over time. AI has many applications in various fields. For example, AI can analyze patient data in healthcare to assist doctors in diagnosis and treatment planning. In finance, AI can predict market trends and improve investment strategies. In transportation, AI can be used to develop autonomous vehicles that can navigate roads and traffic without human intervention. While AI can potentially revolutionize many industries, it raises ethical and societal concerns. For example, there are concerns about the impact of AI on employment, as machines may replace human workers in specific jobs. There are also concerns about bias in AI algorithms, which can lead to unfair treatment of certain groups. In summary, artificial intelligence is a rapidly advancing field that can transform many industries. By creating machines that can learn and adapt to new situations, AI has the potential to improve our lives in countless ways. However, as with any new technology, there are also potential risks and ethical considerations that must be carefully considered. All information regarding artificial intelligence is covered in Chap. 5.

Finally, Chap. 6 gives information on various advanced technologies and tools such as the Internet of Things (IoT), robotics, human-machine interfaces (HMIs), AI software, augmented and virtual reality, blockchain, and cybersecurity. Also, this chapter covers open research problems in different domains such as machine learning, biomedical imaging, robotics, natural language processing, and wireless communications.

Chapter 2
System Identification and Its Applications

In the real world, different systems exist with different characteristics, such as time-invariant and time-varying. The systems can be classified in different ways, such as linear and nonlinear systems, time-variant and time-invariant systems, linear time-variant and linear time-invariant systems, static and dynamic systems, causal and non-causal systems, and stable and unstable systems.

- **Linear and Nonlinear Systems:** A system is linear if it satisfies the following property (in Eq. 2.1), where input signals $A(t)$ and $B(t)$ while output signals $C(t)$ and $D(t)$, respectively. Those systems that are not following this property are known as nonlinear systems.

$$H[A(t) + B(t)] = H[A(t)] + H[B(t)] = C(t) + D(t) \tag{2.1}$$

  In Eq. 2.1, H is the transfer function of the system.
- **Time-Invariant and Time-Variant Systems:** The time-invariant system is a system such that the response doesn't change with time, while the system response changes with time in a time-variant system. The time-invariant systems [1] and time-variant systems are also known as time-in-varying systems and time-varying systems [2].
- **Linear Time-Variant and Linear Time-Invariant Systems:** Linear and time-invariant systems are called linear time-invariant systems or LTI systems. In the field of systems, these systems play a very significant role. Due to the mathematical nature of these systems, any input signal can be analyzed for its output properties. An LTI system can be composed of many LTI systems, resulting in an LTI system in its own right.
- **Static and Dynamic Systems:** A static system is one whose output depends only on its current input. A dynamic system, however, is one whose output depends on its past input.

R. Thanki, P. Joshi, *Advanced Technologies for Industrial Applications*,
https://doi.org/10.1007/978-3-031-33238-8_2

- **Causal and Non-causal Systems:** As static and dynamic systems are distinguished, causal systems are those whose outputs depend only on the present and past state of inputs.
- **Stable and Unstable Systems:** When the outputs and inputs of a system are bound, it is considered stable. An unstable system has a bounded input but an unbounded output.

Various nonlinear, time-varying [18] (variant) systems exist in the real world. The modeling of time-varying nonlinear characteristics and the nonparametric tracking of these properties are well-known and play a significant role in system identification. The weighted least squares method has been employed to identify such systems. This study aims to monitor time-varying nonlinearities while balancing bias and variance using various estimation techniques [3]. Various estimation methods [17] and regression techniques have been used to evaluate how to balance bias and variance.

A modified self-tuning regulator with restricted flexibility has been successfully applied in a large-scale chemical pilot plant. A least squares estimator with the variable weighting of historical data is used in the new approach; at each step, a weighting factor is selected to keep the estimator's scalar measure of information content constant. It is demonstrated that such a method enables the parameter estimations to track gradual and abrupt changes in the plant dynamics for essentially deterministic systems. This chapter has been divided into a few sections and emphasizes the industrial application of time-varying nonparametric systems.

2.1 What Is System Identification?

One definition of a system is an item in which several variables interact throughout a range of time and spatial scales and result in observable signals. Generally, system identification does not have any specific definition but classifies the system to its physical parameters and changes the behavior according to inputs [4].

System identification uses two different data types as a specific statistical inference method. The other, known as a priori, is known before any measurements are made. The first is an experiment. The a priori information [4] broadly refers to the system and signals that are entering the system.

Sometimes it is also classification available based on time-varying or stationary systems. Mostly the classification of system identification is shown in Fig. 2.1. Tracking time-varying systems [5] are the most crucial aspect of adaptive systems, and it is addressed by identifying time-varying processes. Combining linear time-variant and nonparametric methods [6] offers fresh perspectives on how to approach solving statistical issues on their own. Nonlinearities [7] for a time-varying static system have been found using kernel estimation theory [8], orthogonal expansion method, and k-nearest neighbor approach [9]. With prior data knowledge [7] and kernel estimates, the tracking of time-varying nonlinearities can be precisely defined.

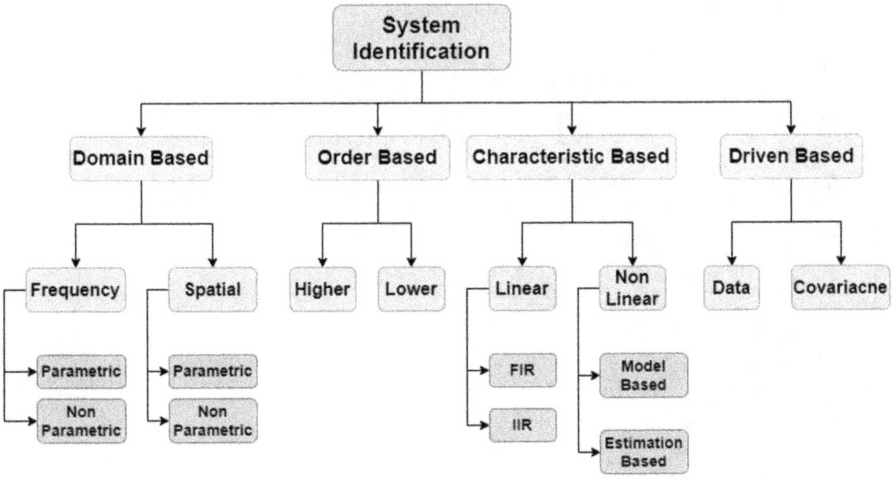

Fig. 2.1 General classification of system identification methods

The nonlinear systems [7] are complex to control and not easy to operate. However, if engineers search for the percentage of nonlinearities, solving the constraint problems can be easy. System identification suggests that the parameter estimation process will become considerably simpler when using the same type of system, specifically for identifying nonlinear systems frequently encountered in applications involving biological or chemical components.

Exercise 1 Let's define one system which can be illustrated as:

$$f(t) = (\alpha_1(x) + \alpha_2(x) + \ldots.. + \alpha_n(x))u_n(t) \tag{2.2}$$

By the above equation, it is clear that:

$$f(t) = \left(\sum_{i=1}^{n} \alpha_i(x)\right) u_n(t) \tag{2.3}$$

For the estimated value, the output should be defined as below:

$$f(y) = \int_0^n f(t)dt = \sum_{i=1}^{n} \alpha_i(x) \int_0^\infty u_n(t)dt \tag{2.4}$$

It is assumed that the inputs must be measured by applying a square to the current point.

$$f(y) = \sum_{i=1}^{n} \alpha_i(x) \int_0^\infty u_n^2(t)dt \tag{2.5}$$

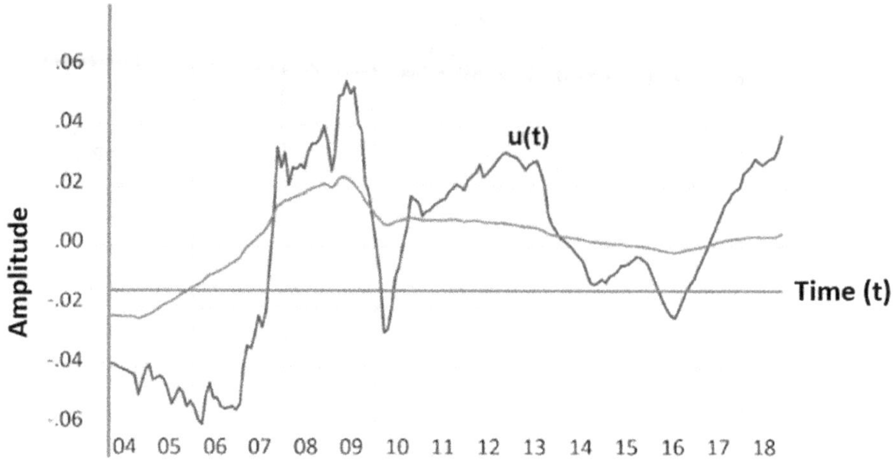

Fig. 2.2 An example of time-varying system

Here, it is clear that, by Fig. 2.2,

$$u_n(t) = E\left|\sum_{i=0}^{n} u_{n-i}(t) - u_0(t)\right| \tag{2.6}$$

Here, it is clear that $f(t)$ and $u_n(t)$ are only measurements and $\alpha_1 + \alpha_2 + ... + \alpha_n$ are sum of the defined parameters. As a result, we cannot independently estimate each parameter from the data.

The main problem in identifiability analysis is the subject of the uniqueness of the estimates, which has garnered a lot of attention in the literature . The analysis [19] is referred to as theoretical, structural, or deterministic identifiability analysis when the query solely considers the situation where the experiment and model structure, in theory, lead to unique parameter values and, therefore, without respect to uncertainties. Most methods for this kind of analysis are only useful for issues with few unknowns. Hence, they are not further studied here.

For instance, the order of data-driven models [10] is sometimes easy to understand but complex to identify based on a percentage of nonlinearity. However, block-oriented models [6] are usually solved using the kernel algorithm and Hammerstein-Wiener models and also support vector machine schemes. The block-oriented parametric [20] and nonparametric system identifications are mentioned in the next section.

2.2 Parametric and Nonparametric System Identification

Prior knowledge plays a crucial role in system identification, comprising the three basic elements of prior knowledge, objectives, and data. It should be understood that these organizations are not autonomous. Data is frequently gathered based on prior system knowledge and modeling goals, resulting in an appropriate experimental design. At the same time, observed data may also cause one to change one's objectives or even one's prior understanding.

The model's structure on physical laws and extra relationships with matching physical parameters is a logical choice at that point, leading to a structure known as a "white-box model." However, if some of these characteristics are unknown or uncertain and, for example, accurate forecasts must be made, the parameters can be inferred from the data. These movable parameters are found in model sets or "gray-box" models. In other situations, such as control applications, linear models are typically adequate and won't always refer to the process's underlying physical laws and relationships. These models are frequently referred to as "black-box" models. Along with selecting the structure, we must also select the model representation, such as the state space, impulse response, or differential equation model representation, and the model parameterization, which pertains to selecting the variable parameters [1].

The identification method, which numerically solves the parameter estimation problem, must be chosen to quantify the fit between model output and observed data. A criterion function must also be supplied. The model's [16] suitability for its intended use is then evaluated in a subsequent phase known as model validation [2, 11]. If the model is deemed suitable at that point, it can be used; otherwise, the method must be repeated, which is typically the case in practice.

2.2.1 Parametric Model Estimation Method

Gray-box models, in which the structure of the dynamics as a function of the parameters is known, but the values of the parameters are unknown, are used in many applications. Due to sensor constraints, these parameters are frequently unavailable and cannot be directly measured.

2.3 Optimization for Time-Varying System

Control theory has a subfield known as optimization for time-varying systems. This subfield focuses on designing and implementing optimal control techniques for time-varying systems whose parameter values, dynamics, or disturbances change with time [5]. The objective of time-varying optimization is to design control policies that either minimize a specified cost function or maximize a specified performance metric over a specified time horizon. This must be done while considering the system's dynamics and constraints on the control inputs and state variables [12].

Time-varying optimization can be approached from various perspectives, such as model predictive control (MPC) [7], adaptive control, and robust control [10]. MPC is a widely used method for time-varying optimization. In this method, the system model is used to predict the system's future behavior over a finite time horizon, and an optimal control policy is computed by solving an optimization problem at each time step. MPC is one of the most widely used methods for time-varying optimization. The optimization problem often comprises a quadratic cost function that makes trade-offs between tracking performance [13], control effort, and other objectives while being subject to constraints on the system states and inputs. This type of cost function is called a "trade-off."

An additional method for time-varying optimization is known as robust control. This method entails the construction of control strategies that are resistant to the effects of uncertainty and disruptions in the system's parameters and dynamics. Another strategy for time-varying optimization is known as adaptive control. This strategy entails modifying the control policy based on online system characteristics and dynamic measurements to achieve optimal results.

In the next sections, a few methods have been discussed with constrained-based time-varying systems.

2.3.1 *Adaptive κ-Nearest Neighbor Method*

As mentioned in Chap. 5, KNN is a nonparametric approach [9]; no underlying assumptions regarding the distribution of the input dataset are necessary. Some prior knowledge of the input dataset is needed to identify the relevant properties. In Fig. 2.2, an example of the time-varying graph has been presented.

A variation of the normal KNN technique that adjusts the value of k to the local characteristics of the data is called the adaptive KNN (κ-nearest neighbor) method [14]. This is accomplished by employing a distance-weighted function, which gives the κ neighbors varying weights based on how close they are to the query point. The following algorithm can be used to define the adaptive KNN technique:

Here the problem is defined for the equation below. Let's consider that we need real-time and live data points w.r.t past data points.

$$u_n(t) = E\left|\sum_{i=0}^{n} u_{n-i}(t) - u_0(t)\right|^2 \tag{2.7}$$

Algorithm 1 An adaptive κ-NN algorithm

Require: Training set u_{n-i}, test instance i, maximum value of κ
Ensure: Predicted class label y for test instance n

$\kappa \leftarrow 1$
while $\kappa \leq max_{\kappa}$ **do**
  Find the κ nearest neighbors of u_t in u_{n-i}
  Calculate the majority class label of the κ neighbors
  if Majority class label of κ neighbors is unique **then**
    return the majority class label
  else
    $\kappa \leftarrow \kappa + 1$
  end if
end while
return random class label

2.3.2 Robust Control Method

A group of control methods known as robust control is created to manage uncertainty and disruptions in a system. When it comes to time-varying systems, where the system characteristics could change over time, robust control methods are especially beneficial. Using adaptive control techniques is one way to identify time-varying systems employing robust control. A group of control methods known as adaptive control modify the controller's parameters according to the system's state at the time. This allows the controller to adjust as the system parameters vary over time.

Model-based control techniques provide a different strategy for time-varying system identification employing robust control. Model-based control entails using a mathematical system model to create the controller. Based on the discrepancy between the actual system behavior and the behavior anticipated by the model, the controller can modify its settings.

The system dynamics and the uncertainties in the system parameters must be carefully considered for both techniques. Also, the requirements of the particular application, such as response time, stability, and robustness to disturbances, must be considered in the controller's design.

Overall, employing robust control approaches can be a successful method for identifying time-varying systems, especially when the system parameters are ambiguous or dynamic. The individual system and application needs must be carefully considered during these strategies' design and implementation.

Constrained-based robust control methods are a class of control techniques that account for system uncertainties and restrictions on the control inputs and system states. These techniques are often employed when there are significant system uncertainties, and it is important to guarantee that the system is stable and operates effectively under all conceivable operating scenarios.

Constrained-based robust control approaches aim to create a controller that can cope with various uncertainties and disruptions. To accomplish this, the

Algorithm 2 Robust control for time-varying system

Require: System model $\dot{y} = f(x, u, t, \alpha)$, control law $u = \mu(x, t, \alpha)$, disturbance bound D

1: Initialize control input $u_0(t)$
2: **for** $k = 0, 1, 2, ...$ **do**
3: Measure state x_k
4: Compute disturbance estimate $\hat{\alpha}k$
5: Compute control input $uk + 1 = \mu(x_k, t_k, \hat{\alpha}k)$
6: Apply control input $uk + 1$ to the system
7: Measure output y_k
8: Compute error $e_k = r_k - y_k$
9: Compute sliding surface $s_k = e_k + D\text{sgn}(e_k)$
10: Compute control law update $D\mu_k = -k_\mu s_k$
11: Update control law $\mu_{k+1} = \mu_k + D\mu_k$
12: **end for**

control problem is often formulated as an optimization problem, with the goal of minimizing a performance metric while considering constraints on the control inputs and system states.

$$f(y) = E\left|\sum_{i=0}^{n} u_{n-i}(t) - u_0(t)\right|^2 \tag{2.8}$$

By the equation above, it can be assumed that:

$$f(y) = \sum_{i=1}^{n} \alpha_i(x) \int_0^\infty E\left|\sum_{i=0}^{n} u_{n-i}(t) - u_0(t)\right|^2 dt \tag{2.9}$$

In this above algorithm, we are making the assumption that the system is described by the differential equation $\dot{y} = f(x, u, t, \alpha)$, where x is the state of the system, u is the control input, t is time, and α represents uncertain system parameters. In other words, we are assuming that the system is described by the differential equation. We also make the assumption that we possess a control rule denoted by the notation $\mu(x, t, \alpha)$ that translates the current state of the system as well as the current time into a control input.

The method employs a sliding mode control strategy to deal with uncertainty in the system parameters. Every time step, we estimate the unknown parameters $\hat{\alpha}_k$ and measure the system state. Then, we compute the control input $uk + 1$ for the subsequent time step using the control rule $\mu(x_k, t_k, \hat{\alpha}k)$.

After applying the control input to the system and measuring the output, we calculate the error e_k between the desired output r_k and the actual output y_k. By computing a sliding surface with the help of this mistake, we can update the control law by applying the formula $D\mu_k = -k_\mu s_k$, where k_μ is a tuning parameter.

Until the required control performance is attained, the algorithm iterates continuously, updating the control law and estimating the uncertain parameters at each time step.

2.4 Industrial Applications of Time-Varying System

Systems that change with time are said to have time-varying properties or behaviors. These systems are used in various industrial applications, and it is essential to analyze and control them to improve and optimize them. These are some examples of time-varying systems being used in the industrial environment:

2.4.1 *Robotic-Based Automotive Industries*

Time-varying systems are widely used in industrial applications, with robotic-based automobile industries being one such use. Automotive manufacturing employs sophisticated robotic systems that must complete various jobs quickly, accurately, and accurately. These robotic systems can be made to work at their best using time-varying systems, which will boost productivity, lower costs, and improve quality control.

The adaptive control system is a time-varying system utilized in the robotic-based automobile industry. Adaptive control systems employ a control algorithm that changes as the environment and the system's behavior do. Adaptive control can be used in robotic systems to modify the control inputs to account for changes in the environment or system dynamics, such as shifting workpiece positions or the presence of outside disturbances.

Trajectory planning is another area where time-varying systems are used in robotic-based automobile sectors. In trajectory planning, the ideal path for a robotic system to complete a particular task is determined. Algorithms for trajectory planning that adjust to changes in the system's environment and behavior over time can be created using time-varying systems.

For instance, a trajectory planning algorithm may adjust the robot's path as it approaches the workpiece based on data from the system's sensors, ensuring that it follows the intended path and avoids running into other items.

Finally, defect finding and diagnostics can leverage time-varying systems. Wear and tear, environmental variables, or other factors may contribute to failures in a robotic-based car manufacturing system. Real-time defect detection and diagnosis are possible with time-varying system approaches, enabling quick correction and minimizing downtime.

2.4.2 Chemical Industries

Time-varying systems are widely used in the chemical industries, where they are utilized to increase the quality of chemical production and optimize process control to achieve maximum efficiency. Time-varying systems can be used to optimize each stage of a chemical process, leading to greater yields, reduced waste, and enhanced profitability. Chemical processes are frequently complex and involve a number of phases. Many procedures (as mentioned below) in the chemical industry can be utilized and handled by time-varying processes:

- **Fault Detection and Diagnosis (FDD):** FDD techniques can detect and diagnose errors in real time, enabling fast corrective action and reducing downtime. FDD techniques are also known as fault tree analysis (FTA). Fault detection and diagnosis (FDD) can use time-varying system methodologies to better account for changes in the process dynamics over time and increase the accuracy of problem detection.
- **Process Design and Optimization:** Time-varying models of the process can be used to simulate the behavior of the process under various operating conditions to optimize the process design to achieve a desired performance objective, such as maximum yield or minimum waste. This can be accomplished through the use of time-varying models of the process.

2.4.3 Communication and Networking

Time-varying systems in communication and navigation have different subcategories as below:

- **Wireless Communication:** Time-varying systems are utilized in wireless communication systems to a significant extent. In these kinds of systems, data is transferred from one place to another by means of electromagnetic waves; these waves change as time passes. To figure out the time-varying signal [15] and retrieve the transmitted data, the receiver must be able to interpret it.
- **Radar Systems:** To identify and locate objects, radar systems transmit and receive signals that fluctuate over time. The radar broadcasts a signal into the air, which then reflects off of the target and is received by the radar. The distance to the object can be determined by the radar by measuring the amount of time that elapses between the signals being broadcast and received. The Doppler effect can also be used to calculate the speed of the studied item.
- **Routing Protocols for Coverage Mobility:** In networking, routing algorithms are used to choose the optimum route for data to take between various network nodes. Shortest Path First (SPF) utilizes Dijkstra's algorithm or a comparable method to determine the shortest path between a source and a destination node.

Each node in the link-state routing (LSR) algorithm keeps a complete map of the network architecture. Nodes frequently share details on their own local linkages and the links of their neighbors. Each node builds a complete map of the network architecture using this data, which it then utilizes to calculate the shortest path between a source and a destination node. Data can be sent from one node in a network to numerous nodes using the routing mechanism known as multicast.

2.4.4 Agriculture and Smart Farming

Time-varying systems can monitor plant development, soil conditions, and environmental elements including temperature, humidity, and light in smart agriculture. Smart agriculture uses closed-loop control systems to maintain greenhouse temperatures. In a closed-loop control system, sensors report greenhouse temperature to a controller. The controller controls the heating or cooling system to maintain the setpoint. Because the greenhouse temperature changes over time, the controller must adjust the heating or cooling system.

Time-varying systems can regulate greenhouse humidity, light, and temperature. A closed-loop control system might measure greenhouse humidity and alter the ventilation system to control air moisture.

2.4.5 Logistics and Storage Industries

Time-varying systems track inventory, optimize warehouse operations, and boost supply chain efficiency in logistics and storage. Real-time inventory tracking is a logistics and storage time-varying system. Sensors are utilized all across a warehouse or supply chain to keep track of the flow of products as part of an inventory tracking system that updates in real time. The location and status of products might change over time as they are transferred from one area to another, which is one of the reasons why the system is considered to be time-varying.

The system monitors the location of goods in real time by employing a number of different sensors, such as radio frequency identification (RFID) tags and barcode scanners. After that, the information is sent to a centralized database or control system, which is then accessible by warehouse managers, supply chain coordinators, and any other relevant stakeholders. In conclusion, time-varying systems have several applications in the logistics and storage industries, such as real-time inventory tracking systems, which can help to optimize warehouse operations and increase supply chain efficiency. These systems also have a number of other uses.

References

1. L. Ljung, Estimating linear time-invariant models of nonlinear time-varying systems. Eur. J. Control **7**(2–3), 203–219 (2001)
2. T. Zhang, W.B. Wu, Time-varying nonlinear regression models: nonparametric estimation and model selection. Inst. Math. Stat. Ann. Stat. **43**(2), 741–768 (2015)
3. M.A.C.I.E.J. Niedzwiecki, First-order tracking properties of weighted least squares estimators. IEEE Trans. Autom. Control **33**(1), 94–96 (1988)
4. M. Gevers, Identification and validation for robust control, in *Advances in Theory and Applications, Iterative Identification and Control* (Springer, London, 2002), pp.185–208
5. P. Joshi, G. Mzyk, Nonparametric tracking for time-varying nonlinearities using the Kernel method, in *New Advances in Dependability of Networks and Systems: Proceedings of the Seventeenth International Conference on Dependability of Computer Systems DepCoS-RELCOMEX, June 27–July 1, 2022, Wrocław, Poland* (Springer International Publishing, Cham, 2022), pp. 79–87
6. G. Mzyk, *Combined Parametric-Nonparametric Identification of Block-Oriented Systems*, vol. 238 (Springer, Berlin, 2014)
7. X. Zhang, J. Liu, X. Xu, S. Yu, H. Chen, Robust learning-based predictive control for discrete-time nonlinear systems with unknown dynamics and state constraints. IEEE Trans. Syst. Man Cyber. Syst. **52**(12), 7314–7327 (2022)
8. M.P. Wand, M.C. Jones, *Kernel Smoothing* (CRC Press, Boca Raton, 1994)
9. G. Biau, L. Devroye, *Lectures on the Nearest Neighbor Method*, vol. 246 (Springer International Publishing, Cham, 2015)
10. A. Nicoletti, A. Karimi, Robust control of systems with sector nonlinearities via convex optimization: a data-driven approach. Int. J. Robust Nonlinear Control **29**(5), 1361–1376 (2019)
11. G. Mzyk, Parametric versus nonparametric approach to Wiener systems identification, in *Block-Oriented Nonlinear System Identification* (Springer, London, 2010), pp. 111–125
12. M. Niedzwiecki, *Identification of Time-Varying Processes* (Wiley, New York, 2000), pp.103–137
13. L.E.S.Z.E.K. Rutkowski, On nonparametric identification with a prediction of time-varying systems. IEEE Trans. Autom. Control **29**(1), 58–60 (1984)
14. S. Sun, R.Huang, An adaptive k-nearest neighbor algorithm, in *2010 Seventh International Conference on Fuzzy Systems and Knowledge Discovery*, vol. 1 (IEEE, Piscataway, 2010), pp. 91–94
15. V. Ingle, S. Kogon, D. Manolakis, *Statistical and Adaptive Signal Processing* (Artech, London, 2005)
16. W. Greblicki, M. Pawlak, *Nonparametric System Identification*, vol. 1 (Cambridge University Press, Cambridge, 2008)
17. L.A. Liporace, Linear estimation of nonstationary signals. J. Acoust. Soc. Amer. **58**(6), 1288–1295 (1975)
18. L. Rutkowski, On-line identification of time-varying systems by nonparametric techniques. IEEE Trans. Autom. Control **27**(1), 228–230 (1982)
19. M.J. Niedźwiecki, M. Ciołek, A. Gańcza, A new look at the statistical identification of nonstationary systems. Automatica **118**, 109037 (2020)
20. F. Giri, E.W. Bai (eds.), *Block-Oriented Nonlinear System Identification*, vol. 1 (Springer, London, 2010), pp.0278–0046

Chapter 3
Signal Processing and Its Applications

Everything we use and rely on in our daily lives is enabled by signal processing. Signal processing is a branch of electrical engineering that analyzes data generated using physical devices using various models and theories. It models and analyzes data representations of physical events and data generated across multiple disciplines. These devices include computers, radios, video devices, cellphones, intelligent connected devices, and much more. Our modern world relies heavily on signal processing. This field combines biotechnology, entertainment, and social interaction. Our ability to communicate and share information is enhanced by it. We live in a digital world thanks to signal processing.

Signal processing refers to any modification or analysis of a signal. These processing techniques are used to improve the efficiency of the system. Signal processing has applications in nearly every field of life. But, before we get into that, let us define signal. A signal is an electrical impulse or a wave that carries information. The electrical impulse refers to the changing currents, voltages, or electromagnetic waves that transmit data at any point in electrical systems. Examples of signals are speech, voice, video stream, and mobile phone signals. The noise is also considered a signal, but the information carried by noise is unwanted; that is why it is considered undesirable. Let us briefly go through its types.

3.1 Basic of Signal Processing

This section briefly discusses basic information about signal processing; the mathematics used in signal processing, particularly various transforms; etc.

R. Thanki, P. Joshi, *Advanced Technologies for Industrial Applications*,
https://doi.org/10.1007/978-3-031-33238-8_3

3.1.1 Types of Signal Processing

Signal processing is classified into different categories based on the types of signals. These categories are analog signal processing, digital signal processing, nonlinear signal processing, and statistical signal processing. The information of basic types of signals are as per below:

- **Analog Signal Processing:** A continuous signal has not been digitized. In this case, its values are typically represented as a voltage, an electric current, or an electrical charge around components. There are many applications in the real world where analog signal processing is still relevant, and even when sampling and discretizing signals for digital processing, it is still the first step.
- **Digital Signal Processing:** Signals that have been digitized and sampled discretely in time. Digital circuits such as specialized digital signal processors (DSPs), FPGAs, or ARM chips perform processing to convert an analog signal into a digital version [1]. In many applications, digital processing offers several advantages over analog processing, such as error detection and correction, data compression, and error correction in transmission. Digital wireless communication and navigation systems are also based on this technology.
- **Nonlinear Signal Processing:** Because linear methods and systems are easy to interpret and implement, classical signal processing relies on linear methods and techniques. Some applications, however, would benefit from nonlinear processing methods being included in the methodology. Several nonlinear signal processing methods have proven efficient in addressing real-world challenges, including wavelet and filterbank denoising, sparse sampling, and fractional processes.
- **Statistical Signal Processing:** Modeling the system under study is often beneficial for many applications. However, unlike physical models such as a swinging pendulum, it is impossible to predict the behavior of most signals of interest with 100% accuracy. It will be necessary to include as many "broad" properties as possible, such as the variation and correlation structure, to develop a model for such a signal. Mathematics and stochastic processes are best used to describe this phenomenon. It is possible to express optimality criteria and evaluate achievable performance using these models.

3.1.2 Types of Different Systems

The systems can be classified in different ways such as linear and nonlinear systems, time-variant and time-invariant systems, linear time-variant and linear time-invariant systems, static and dynamic systems, causal and non-causal systems, and stable and unstable systems. The system can be analyzed using various signals, such as analog and digital.

Transmission of information (including audio and video) is usually carried out via analog or digital signals. Analog technology translates information into electric pulses of varying amplitudes, while digital technology converts it into binary format (either 0 or 1). The analog and digital systems are called continuous- and discrete-time systems. Due to the better utilization of various parameters such as power, memory, hardware, and cost in digital systems compared to analog systems, it is widely used nowadays. Now onward, we are discussing digital or discrete-time systems in detail in terms of their analysis methods and various usages of them in various industries.

An electronic system that uses discrete-time signals to operate is called a discrete-time or digital system. Signals with square waveforms are used in digital systems. Using a sampling technique, digital systems transform analog signals into digital ones. The system produces a desired output once the equivalent digital signal has been produced. Compared to analog systems, digital systems are slower because of this conversion process. However, digital systems have several advantages, including noise-free data transmission, efficiency, ease of implementation, cost-effectiveness, reliability, etc. As a result of all these advantages, digital systems are becoming more popular.

3.2 Transforms Used for Analysis of Signals and Systems

An output signal is generated by transforming the original signal into the resulting signal, which is a mathematical model. A complex operation can be decomposed into a sequence of simpler ones, although it is often convenient to describe a complicated operation using this concept. According to the Unified Signal Theory, the output domain can differ from the input domain. Additionally, multidimensional signals can be transformed between different dimensions. Various transforms such as Laplace, Z, Fourier, and wavelet are used to analyze various kinds of systems where inputs and outputs are some kinds of signals.

3.2.1 Laplace Transform

In 1980, Laplace proposed the Laplace transform (LT). This operator transforms signals in the time domain into signals in a complex frequency domain called the "S" domain. In this case, "S" represents the complex frequency domain, and "*s*" represents the complex frequency variable. The complex frequency S can be likewise defined as:

$$s = \sigma + j\omega \tag{3.1}$$

where σ is the real part of s and $j\omega$ is the imaginary part of s.

Mathematicians define complex numbers as mathematical abstractions used for analyzing signals and systems. It simplifies mathematics. In the same way, complex frequency planes are also useful abstractions for simplifying mathematics. Except for the fact that it converts a time-domain signal into a complex frequency-domain signal, the Laplace transform has no physical significance. Signals and systems can be easily analyzed using it for simplifying mathematical computations. Knowing the Laplace domain transfer function of a system reveals its stability directly. Differential equations can be solved using LT.

3.2.2 *Z-Transform*

In mathematical terms, the Z-transform converts difference equations from the time domain to the algebraic domain. Z-transforms are very useful tools for analyzing linear shift-invariant systems (LSIs). Different equations are used to represent LSI discrete-time systems. To solve these time-domain difference equations, the Z-transform is first used to convert them into algebraic equations in the z-domain. Then the algebraic equations are manipulated in the z-domain, and then the result is converted back into the time domain by using the inverse Z-transform. There are two types of Z-transform: unilateral (one-sided) and bilateral (two-sided).

Mathematically, if $x(n)$ is a discrete-time signal or sequence, then its bilateral or two-sided Z-transform is defined as:

$$Z[x(n)] = X(z) = \sum_{n=-\infty}^{\infty} x(n)z^{-n} \tag{3.2}$$

A complex variable z can be expressed as follows:

$$z = r \cdot e^{j\omega} \tag{3.3}$$

In Eq. 3.3, a circle's radius is defined by r. In addition, the unilateral Z-transform is defined as follows:

$$Z[x(n)] = X(z) = \sum_{n=0}^{\infty} x(n)z^{-n} \tag{3.4}$$

One-sided or unilateral Z-transforms are very useful when dealing with causal sequences. Furthermore, it is primarily used to solve differential equations with initial conditions. It is called the region of convergence (ROC) of the Z-transform $X(z)$ when the Z-transform of a discrete-time sequence $x(n)$ converges for a set of points in the Z-plane. It is possible for the Z-transform to converge or not for any discrete-time sequence. The sequence $x(n)$ has no Z-transform if the function $X(z)$ does not converge in the Z-plane. The Z-transform has the following advantages:

- A discrete-time system can be analyzed more easily with the Z-transform by converting its difference equations into simple linear algebraic equations.
- In the z-domain, convolution is converted to multiplication.
- There is a Z-transform for signals that cannot be transformed by the discrete-time Fourier transform (DTFT).

Z-transforms have the primary disadvantage of not being able to obtain the frequency-domain response and plot it.

3.2.3 Fourier Transform

Fourier transforms decompose waveforms, or functions of time, into their frequencies. Fourier transforms produce complex-valued functions of frequency. Fourier transforms represent the frequency values present in the original function, and their complex arguments represent the phase offset of the sinusoidal. It is also known as a generalization of the Fourier series. This term encompasses both a mathematical function and a frequency-domain representation. With the Fourier transform, any function can be viewed as a sum of simple sinusoids, which allows the extension of the Fourier series to non-periodic functions. The Fourier transform of a function $x(k)$ is given by:

$$X(k) = \sum_{n=0}^{N-1} x(n) \cdot e^{-2j\pi nk} \tag{3.5}$$

where $x(k)$ is the input signal in a time domain and $X(k)$ is the transformed signal in the frequency domain.

Fourier transforms have the following properties:

- This is a linear transformation. In this example, we can calculate the Fourier transform of the linear combination of a and b if $a(k)$ and $b(k)$ are two Fourier transforms given by $A(k)$ and $B(k)$.
- Timeshift is one of its properties. As a result of the Fourier transform of $x(t–a)$, the magnitude of the spectrum is also shifted by the same amount as the shift in the original function.
- It has the property of modulation. When a function is multiplied by another function, it is modulated by that function.
- Parseval's theorem is used in its formulation. Fourier transforms are unitary, so a function $a(k)$'s square root equals its Fourier transform $A(k)$.
- It has the property of duality. The Fourier transform of $a(k)$ is $A(-k)$ if $a(k)$ has the Fourier transform $A(k)$.

There are two types of Fourier transform: discrete-time or discrete Fourier transform (DTFT or DFT) and fast Fourier transform (FFT).

3.2.3.1 Discrete Fourier Transform (DFT)

Mathematically, the discrete Fourier transform (DFT) transforms an equally spaced sequence of samples of a function into an equally spaced sequence of discrete-time Fourier transforms (DTFT), which are complex-valued functions of frequency. Digital signal processing relies heavily on the discrete Fourier transform (DFT). Frequency-domain (spectral) representations of signals are derived from them.

The discrete Fourier transform is very similar to the Fourier transform in mathematics. Specifically, given a vector of n input amplitudes such as f_0, f_1, $f_2, \ldots, f_{n-2}$, f_{n-1}, the discrete Fourier transform yields a set of n frequency magnitudes. The DFT is defined as such:

$$X[k] = \sum_{n=0}^{N-1} x[n] e^{\frac{-j2\pi kn}{N}} \tag{3.6}$$

where $X(k)$ is used to denote the Fourier transformed signal, $x[n]$ is used to denote the original signal, and N is the length of the sequence to be transformed.

The inverse DFT is defined as such:

$$x[n] = \frac{1}{N} \sum_{k=0}^{N-1} X[k] W_N^{-kn} \tag{3.7}$$

where W_N is defined as:

$$W_N = e^{\frac{-j2\pi}{N}} \tag{3.8}$$

3.2.3.2 Fast Fourier Transform (FFT)

Fourier transforms can be generated more efficiently using the fast Fourier transform (FFT). FFT's main advantage is speed, which reduces the number of calculations required to analyze a waveform. In addition, it is used to design electrical circuits, solve differential equations, process signals, analyze signals, and filter images.

3.2.4 Wavelet Transform

We can use wavelets to extract more useful information from any signal by transforming it from one representation to another. It is known as a wavelet transform. Wavelet transforms can be mathematically represented as convolutions between wavelet functions and signals. Signals in time-frequency space can be analyzed with the wavelet transform (WT) to reduce noise while preserving significant components. Signal processing has benefited greatly from WT in the past 20 years.

With wavelet analysis, mathematics, physics, and engineering problems are solved in an exciting new way. Wave propagation, data compression, signal processing, image processing, pattern recognition, computer graphics, aircraft and submarine detection, and other medical imaging applications are some of the many applications of wavelet analysis. Composing complex information into elementary forms, such as music, speech, images, and patterns, can be reconstructed with high precision using wavelets.

3.3 Designing of Discrete-Time Systems

Software or hardware implementations of linear time-invariant discrete-time systems are described in this section. This popular class of linear time-invariant discrete-time systems is defined by the general linear constant-coefficient difference equation.

$$b(n) = -\sum_{k=1}^{N} p_k b(n-k) + \sum_{k=0}^{M} q_k a(n-k) \tag{3.9}$$

Z-transforms and the rational system function also describe linear time-invariant discrete-time systems as below:

$$H(z) = \frac{\sum_{k=1}^{M} q_k z^{-k}}{1 + \sum_{k=1}^{N} p_k z^{-k}} \tag{3.10}$$

It is possible to convert the equations obtained by rearranging (3.9) into a program that runs on a computer if the system is to be implemented as software. A block diagram implies a hardware configuration for implementing the system.

For the system to be designed, various factors must be considered, including computational complexity, memory requirements, and finite word length effects. Complicated systems require more arithmetic operations to compute an output value. Inputs, outputs, and any intermediate computed values are stored due to memory requirements. The term finite word length effect refers to quantization effects that are inherent to all digital implementations of the system, both hardware and software.

Three major factors influence our choice of implementing a system of the type described in Eqs. 3.9 and 3.10. However, other factors may also play a role in determining which implementation to use, such as whether the structure or realization lends itself to parallel processing or whether the computations can be pipelined. Digital signal processing algorithms are usually more complex when these additional factors are considered. Any discrete-time system can be designed or realized using two types of filters: finite impulse response (FIR) and infinite impulse response (IIR).

3.3.1 Finite Impulse Response (FIR) Filter

The general equation for the FIR filter can be given below:

$$b(n) = \sum_{k=0}^{M-1} p_k a(n-k) \tag{3.11}$$

or equivalently by system function:

$$H(z) = \sum_{k=0}^{M-1} p_k z^{-k} \tag{3.12}$$

Further, the unit sample response of the FIR filter can be described below:

$$h(n) = \begin{cases} p_n, 0 \leqslant n \leqslant M-1 \\ 0, \text{otherwise} \end{cases} \tag{3.13}$$

The length of the FIR filter is set to M. The direct method is a simple structure used in the literature for implementing a FIR system [2]. The FIR filter can be realized using different structures such as cascades, frequency sampling, and lattices. The following are the primary advantages of FIR filters:

- A linear phase can be achieved by them.
- It is always stable with them.
- There is generally a linear approach to design.
- It is feasible to implement them in hardware efficiently.
- There is a finite duration to the filter startup transients.

A major disadvantage of FIR filters is that they typically require much higher filter orders to achieve the same performance levels as IIR filters. Consequently, these filters are often much slower than IIR filters with equal performance. The FIR filter can be designed using various methods such as windowing, multiband with transition bands, constrained least squares, arbitrary response, and raised cosine [3].

3.3.2 Infinite Impulse Response (IIR) Filter

A system described by Eqs. 3.11 and 3.12 can be realized using an IIR system using direct-form, cascade, lattice, and lattice-ladder structures similar to FIR filters. One difference is that the IIR filter is realized parallel rather than serially in the FIR filter [2]. An IIR filter is generally more cost-effective than a corresponding FIR filter since it meets a set of specifications with a much lower filter order [4]. The IIR filter can be designed using various methods such as analog prototyping, direct design, generalized Butterworth design, and parametric modeling [4]. The IIR filter types such as classical IIR filters, Butterworth, Chebyshev Types I and II, elliptic, and Bessel are available in the literature [4].

3.4 Industrial Applications of Signal Processing (SP)

The IEEE Signal Processing Society's Industry Digital Signal Processing (DSP) Standing Committee (IDSP-SC) focuses on identifying and evaluating emerging digital signal processing applications and technologies [5]. There are several signal processing applications and technologies recommended by the committee, including digital and software radio frequency (RF) processing, single-chip solutions, nanoscale technology, cognitive and reconfigurable radar, the Internet of Things, cloud computing, service computing, and new-generation TV (smart TV, 3D TV, 4K TV, UHD TV), and perception by autonomous systems [5].

3.4.1 SP for Digital Front End and Radio Frequency

Radar, sonar, digital broadcasting, and wireless communication rely heavily on software and digital processing at the front end. These techniques offer low power consumption, low costs, fast time to market, and flexibility. Unlike baseband processing, the front end is tightly connected to the RF layer, which imposes significant limitations and difficulties on digital processing speed, memory, computational capability, power, size, data interfaces, and bandwidths. This suggests that digital processing and circuit implementation of the front end are challenging tasks that require huge efforts from industry, research, and regulatory authorities.

3.4.2 Development of Chip for All DSP

Signal processing algorithms have been converted to silicon using three different computing platforms: application-specific integrated circuits (ASICs), digital signal processors (DSPs), and field-programmable gate arrays (FPGAs). A single application device usually incorporates a variety of signal processing algorithms. This suggests that this single application device needs different computing platforms/IC chips, which is practically inefficient.

An ASIC-based solution's power consumption and performance are excellent; however, this solution cannot support multiple standards and applications. The performance of digital processing systems is highly dependent on signal processing algorithms, which cannot be upgraded using an ASIC-based solution. The flexible nature of FPGA- and DSP-based solutions allows them to meet several standards (or models or applications) and support a wide range of signal processing algorithms. However, FPGA/DSP-based solutions could be more efficient from a power consumption and cost perspective. In some cases, these two solutions can be combined and viewed as an accelerator-based platform, producing some advantages in both performance and flexibility. This third solution, however, has a significant problem in that it is difficult to program/port different algorithms into its platform, primarily because its control units, computational units, data units, and accelerators have heterogeneous interfaces. A single-chip solution is highly desirable by combining power efficiency, cost reduction, time to market, flexibility, and programming ability.

3.4.3 Usage of SP in Nanotechnology

Many technologies are in the nanoscale area, including nanonetworks, nanorobotics, nanosecond processors (GHz scale), nanoscale CMOS circuits and sensors, and three-dimensional (3D) integrated circuits. Research and applications in signal processing will be greatly impacted by nanotechnology. Several signal processing algorithms (like matrix inversion in MIMO communication systems) can be performed in nanoseconds when using a processor with a nanosecond instruction period or clock cycle time. Both consumer electronics and military instruments use CMOS-based nanoscale image sensors and related processing to offer much better image systems.

The most significant benefit of nanonetworks is that they are capable of computing, data storage, sensing, and actuating specific tasks because they are interconnected micromachines or devices. Nanonetworks enable the development of more complex systems, such as nanorobots and computing devices incorporating nanoprocessors, nano-memory, or nano-clocks. Nevertheless, signal processing (coding, transmission, implementation) needs to be rethought and redesigned since nanonetworks differ from traditional communication networks in a number of ways, including signal type, coding and the message carried, propagation speed, noise sources, and a limit on a nanoscale size, complexity, and power consumption, among others.

3.4.4 Development of Reconfigurable and Cognitive Radar

An adaptive scheduler, adaptive data product generation, adaptive transmit and receive chain, and enhanced real-time adaptability enable reconfigurable and cognitive radar to adapt intelligently to its environment based on many potential information sources. Research and development of reconfigurable and cognitive radars are related to two signal processing aspects.

Several aspects include adaptive power allocation, knowledge-aided processing, learning, environmental dynamic databases, and data mining. Differentiation technology can be applied across time, frequency, spatial, and embedded domains. Various radar degree-of-freedom (DoF) and channel numbers can be optimized for RF digitization, processing, and digital arbitrary waveform generators (DAWG). A high-performance computing platform is also necessary to implement cognitive radar in real time and reconfigurable. As an example, a computing platform of this type should be not only able to perform all the computations required by the various algorithms for estimating adaptive channel parameters (such as eigenvalue decomposition, matrix inversion, and QR decomposition) in real time but also be able to take advantage of all information sources through knowledge-aided coprocessing capabilities.

3.4.5 SP in Smart Internet of Things

A wide range of devices and places are expected to become IP-enabled and be integrated into the Internet soon. Various examples of intelligent objects include mobile phones, personal health devices, appliances, home security, and entertainment systems. In addition, there are RFID, industrial automation, smart metering, and environmental monitoring systems. There are many benefits that the intelligent Internet of Things can offer. These include environmental monitoring, energy savings, intelligent transportation, more efficient factories, better logistics, smart agriculture, food safety, and better healthcare.

The following related areas will greatly depend on signal processing technology and practice: wireless embedded technology, ubiquitous information acquisition and sensing, RFID algorithms and circuit integration, signal and data coding and compression, security authentication, key management algorithms, and routing algorithms. In the smart grid, a number of significant components are involved in the signal processing process: bulk generation, transmission, distribution, customers, operations, markets, and service providers. Three layers are included: a power and energy layer, a communication layer, and an information technology/computer layer. There is no doubt that signal processing will be primarily used in the second layer, encompassing smart metering and its wireless communication architecture, microcontrollers with ultralow power consumption, models for power grid data and state estimation, and algorithms for fault detection, isolation, recovery, and load balancing in real time.

3.4.6 SP for Cloud and Service Computing

A cloud computing service is a method of computing and processing that utilizes virtualized, dynamic resources (software, multimedia, data access, and storage services) without the need for end users to know the physical location or reconfigure the systems for delivering the services from a business or information service standpoint. Dynamic allocation of cloud resources is essential to maximize the system's performance. Therefore, designing and implementing dynamic resource allocation algorithms will be a crucial signal processing topic in cloud computing. Among the issues discussed are algorithms and real-time implementations for compression, coding, storage, processing, security, privacy, IP management, communication, streaming (ultrahigh bandwidth), modeling, and evaluating the quality of services and experiences [5].

3.4.7 SP for Digital TV Technology

From sensing to transmission to display, digital TV systems and services involve signal processing at every stage. Several signal processing topics are discussed, including digital broadcasting baseband processing, white space, and dynamic spectral management, embedded SoC implementation, cross-layer coding, multi-viewer coding, recording and tracking, representation and segmentation, display technology and color formats, SDR and broadcasting, and human factors and perceptual quality assessment, along with algorithms for managing electronic copyright.

3.4.8 SP for Autonomous System Perception

The long-term goal of machine perception is to allow machines to understand objects, events, and processes in the environment and communicate this understanding to humans. There is a trend to integrate multiple input methods into machine perception. This includes sensors such as radar and microphone arrays. However, machine perception was often synonymous with machine vision, which processes data from cameras operating in the visible range. Currently, sensors are being used as the base layer in a layered model, which includes front-end processing, object localization, object recognition, context recognition, and spatiotemporal perception.

Among the main technical challenges are (1) constructing robust real-world algorithms for mid-level tasks, (2) generating "complete" ontologies of scenes/scenarios of interest, and (3) identifying and describing events beyond the trained set [5]. The challenges associated with autonomous systems are computation, scalability across robot platforms, interfacing with machine intelligence, and human-robot interaction. Multichannel processing of multimodal sensor outputs, cueing and behavior inference, and symbolic representations are among the signal processing problems presented by the field.

References

1. R.C. Gonzalez, R.E. Woods, *Digital Image Processing* (Pearson Education India, Upper Saddle River, 2008)
2. J.G. Proakis, D.G. Manolakis, *Digital Signal Processing: Principles, Algorithms, and Applications* (Pearson Education India, Noida, 2007)
3. FIR Filter Design. Web Link: https://www.mathworks.com/help/signal/ug/fir-filter-design.html. Last Access February 2023
4. IIR Filter Design. Web link: https://www.mathworks.com/help/signal/ug/iir-filter-design.html. Last Access February 2023
5. F.L. Luo, W. Williams, R.M. Rao, R. Narasimha, M.J. Montpetit, Trends in signal processing applications and industry technology [in the spotlight]. IEEE Signal Proc. Mag, **29**(1), 184–174 (2011)

Chapter 4
Image Processing and Its Applications

Whenever we look at a digital image, we see many elements, each with a specific location and value [1]. Pixels, picture elements, and image elements are examples of these elements. Digital images are commonly represented by pixels. What happens when we look at an object? The process begins with the eye capturing the object and sending signals to the brain. The brain decodes these signals and obtains valuable information. Image processing is the process of converting images into useful data.

We begin processing images as soon as we are born and continue doing so until the end, which is an integral part of our lives. Therefore, combining the eye and the brain creates the ultimate imaging system. In image processing, algorithms are written to process images captured by a camera. Here the camera replaces the eye, and the computer does the brain's work. Image processing involves changing the nature of an image to either (1) improve its visual information for human interpretation or (2) make it more suitable for autonomous machine perception.

Today, image processing is used around the world. Image processing applications can be classified based on the energy source used to generate images. The principal energy source for images today is the electromagnetic energy spectrum, and other energy sources may be acoustic, ultrasonic, and electronic [1]. Figure 4.1 shows the electromagnetic energy spectrum. For example, the image generated by gamma-ray is called gamma-ray imaging. The image created by an X-ray is called an X-ray image. These images are widely used in medical science to inspect the human body. Gamma-ray imaging is primarily used in nuclear medicine and astronomy.

A single image can be processed using image processing at one end and viewed through computer vision at the other end. There are three basic types of image processing:

- **Low-Level Image Processing:** Basic operations such as noise reduction, contrast enhancement, and sharpening are included in low-level image processing. These processes use images as inputs and outputs.
- **Medium-Level Image Processing:** Object classification, image segmentation, and description of objects presented in an image are operations included in

R. Thanki, P. Joshi, *Advanced Technologies for Industrial Applications*,
https://doi.org/10.1007/978-3-031-33238-8_4

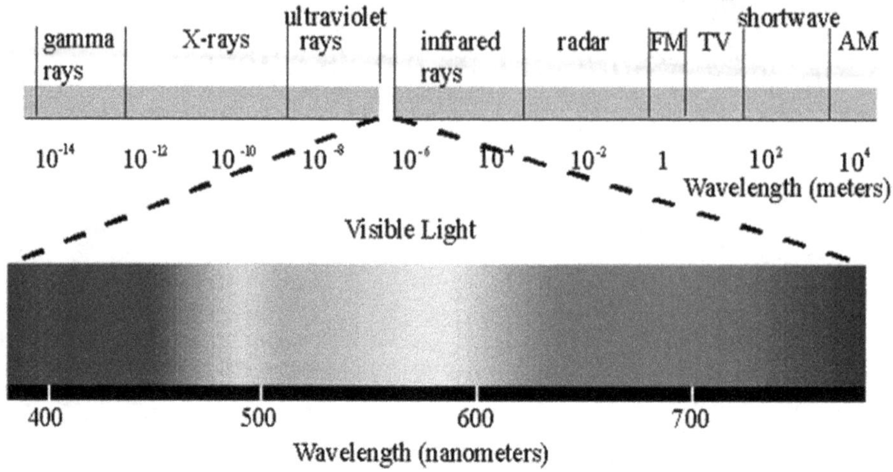

Fig. 4.1 Electromagnetic energy spectrum

medium-level image processing. This process utilizes images as inputs and extracts edges and contours from these images as outputs.

- **High-Level Image Processing:** Analyzing images is part of this process. Images' features are inputs to this process, and images' features are outputs.

In basic image processing steps, images are acquired, enhanced, restored, processed in color, processed with wavelets, compressed, morphologically analyzed, segmented, represented, and described, and objects are recognized. A processed image of a sensed object is acquired as the first step in image processing. Image processing can be applied to images as inputs or to the attributes of images as outputs.

4.1 Most Commonly Used Images

An image is interpreted as a 2D or 3D matrix by a computer, where each pixel represents the amplitude of that pixel or its "intensity." We generally deal with 8-bit images, with amplitude values ranging from 0 to 255. A computer perceives an image as a function $I(x, y)$ or $I(x, y, z)$, where "I" represents the intensity of a pixel and (x, y) or (x, y, z) represents its coordinates (for binary/grayscale or RGB images, respectively). A computer deals with images in different ways depending on their function representation. Let's discuss it. Here we give information on the most commonly used images in the real world. The images may be binary, grayscale, or color images.

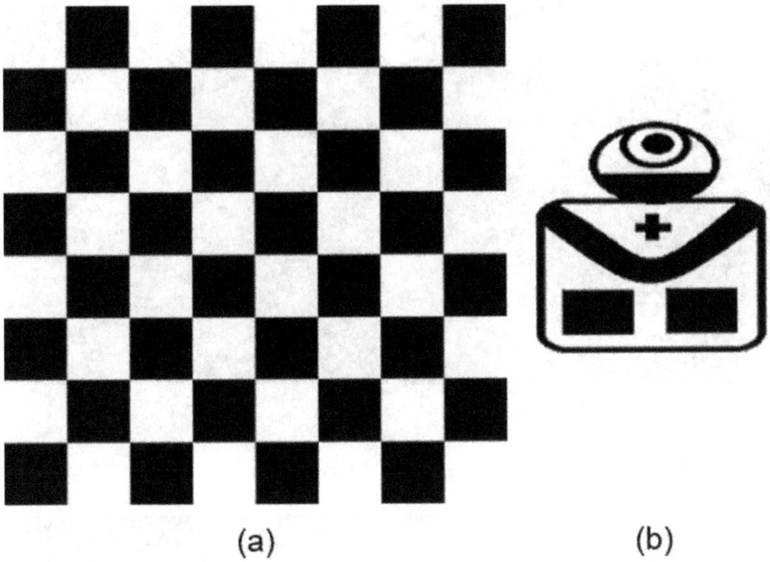

Fig. 4.2 Examples of binary images. (**a**) Chessboard. (**b**) Logo

4.1.1 Binary Image

The image has only two intensity levels, such as 0 for black and 1 or 255 for white, which is called a binary image. This image is widely used in image segmentation and highlights certain regions in a color image. The examples of binary images are shown in Fig. 4.2.

4.1.2 Grayscale Image

An 8-bit grayscale image comprises 256 unique colors, with 0 corresponding to black and 255 representing white. The other 254 values represent shades of gray in between. This image is widely used in most image processing methods. The examples of grayscale images are shown in Fig. 4.3 [2].

4.1.3 Color Image

In our modern world, we are used to seeing RGB or colored images that are 16-bit matrices. Each pixel can have 65,536 different colors. RGB refers to an image's red, green, and blue channels. We used to have images with a single channel up until now.

Fig. 4.3 Examples of grayscale images. (**a**) Lena. (**b**) Rose

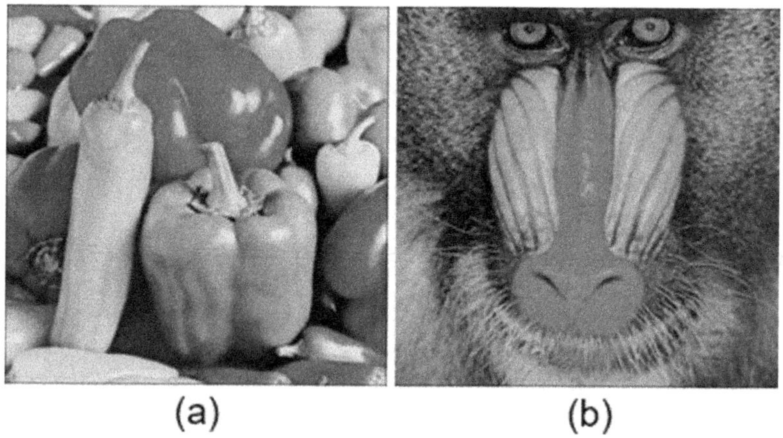

Fig. 4.4 Examples of color images. (**a**) Peppers. (**b**) Baboon

In other words, any value of a matrix could be defined by two coordinates. However, to specify the value of a matrix element, we require three unique coordinates for three equal-sized matrices (called channels), each with a value between 0 and 255. When a pixel value is (0, 0, 0) in an RGB image, it is black. It is white when it is (255, 255, 255). Any combination of numbers between those two can create all the colors in nature. For example, (255, 0, 0) corresponds to red (since only the red channel is active here). The colors (0, 255, 0) and (0, 0, 255) are green and blue, respectively. The examples of grayscale images are shown in Fig. 4.4 [2].

4.2 Fundamental Steps of Image Processing

The fundamental steps of image processing are described as follows [1]:

4.2.1 Image Acquisition

Cameras capture images, which an analog-to-digital converter digitizes (if not automatically digitized).

4.2.2 Image Enhancement

The acquired image is manipulated in this step to meet the specific requirements of the task for which it is intended. Usually, these techniques highlight hidden or significant details in an image, such as adjusting contrast and brightness. The process of image enhancement is highly subjective.

4.2.3 Image Restoration

In this step, the appearance of an image is improved, and a mathematical or probabilistic model is used to explain the degradation. An example would be removing noise from an image or blurring it.

4.2.4 Color Image Processing

During this step, colored images are processed, such as color correction or modeling.

4.2.5 Wavelets and Multi-Resolution Processing

A wavelet is a unit for representing images with different levels of resolution. For data compression and pyramidal representation, images are subdivided into smaller regions.

4.2.6 Image Compression

Images must be compressed to be transferred between devices or accommodate computational and storage constraints. Images are also highly compressed when displayed online; for example, Google provides thumbnails of highly compressed versions of the originals. Images are shown in their original resolution only when you click on them. In this way, the servers can save bandwidth.

4.2.7 Morphological Processing

Image components must be extracted for processing or downstream applications to represent and describe shapes. The morphological process gives us the tools to accomplish this (which are mathematical operations). Sharpening and blurring the edges of objects in an image are achieved using erosion and dilation operations.

4.2.8 Image Segmentation

This step divides an image into different parts to simplify and/or make it easier to analyze and interpret. As a result of image segmentation, computers can focus their attention on the important parts of an image, thereby improving the performance of automated systems.

4.2.9 Representation and Description

This step of the image segmentation procedure involves determining whether the segmented region should be displayed as a boundary or a complete region. The purpose of the description is to extract attributes that provide some quantitative information of interest or can be used to differentiate one class of objects from another.

4.2.10 Object Detection and Recognition

As soon as the objects have been segmented from an image, the automated system needs to assign a label to the object that humans can use to understand what the object is.

4.2.11 Knowledge Base

The information in images relevant to knowledge is highlighted using some methods, such as finding the bounding box coordinates. Anything relevant to solving the problem can be encoded in a knowledge base.

4.3 Image Processing Methods

In image processing, unwanted objects can be removed from an image or even completely different images can be created. A person's picture can be rendered in the foreground using image processing to remove the background. There are a variety of algorithms and techniques that can be used in image processing to achieve a variety of different results. The purpose of this section is to describe different image processing methods [3].

4.3.1 Image Enhancement

The task of image enhancement, or improving an image's quality, is one of the most common tasks in image processing. It plays a crucial role in computer vision, remote sensing, and surveillance. The contrast is typically adjusted to make the image appear brighter and more contrasted. Contrast refers to the difference between an image's brightest and darkest parts. An image can be more visible by increasing its contrast, which increases its brightness. In an image, brightness refers to how light or dark it is. Images can be made brighter by increasing brightness, making them easier to see. Most image editing software allows you to adjust contrast and brightness automatically, or you can do it manually. Image enhancement is performed in two domains such as the spatial domain and the transform domain.

4.3.1.1 Image Enhancement in Spatial Domain

In the spatial domain, images are represented by pixels. Spatial domain methods process images directly based on pixel values. A general equation can be applied to all spatial domain methods.

$$g(x, y) = P(f(x, y)) \tag{4.1}$$

An input image is $f\ (x, y)$, a processed image is $g\ (x, y)$, and a processing operation is P. Pixels (x, y) within the neighborhood of (x, y) are generally considered neighborhood pixels. Each position in the sub-image is processed by

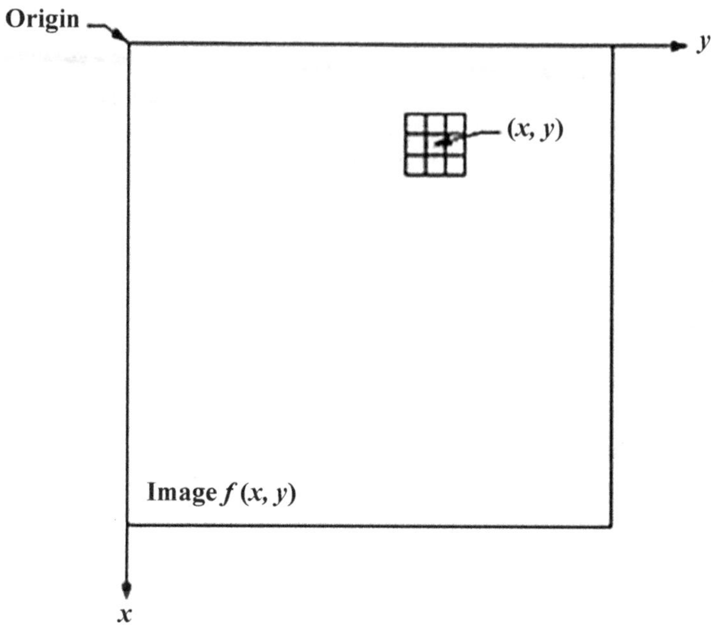

Fig. 4.5 Point processing

applying the processing operation P at each point position to generate processed points. Figure 4.5 shows image processing in the spatial domain. This processing is divided into two types: point processing and neighborhood processing [1, 4].

- **Point Processing:** It is the simplest form of spatial image processing. In other words, it is a transformation of gray levels. Here, P is 1×1, which means that the value of $f\ (x, y)$ depends only on the original value of $g\ (x, y)$. As a result, P becomes a gray-level transformation function.

$$S = P(R) \tag{4.2}$$

  In Eq. (4.2), S is the gray level of the processed image $g\ (x, y)$. R represents the gray level of the original image $f\ (x, y)$. The common operations for this processing are identity transformation, image negative, contrast stretching, contrast thresholding, gray-level slicing, bit plane slicing, log transformation, power law transformation, and histogram processing [1].
- **Neighborhood Processing:** Neighborhood processing extends level transformation by applying an operating function to a neighborhood pixel of every target pixel. The mask process is used in this process. This method will create a new image with pixel values based on the gray-level values under the mask. Figure 4.6 shows this process.

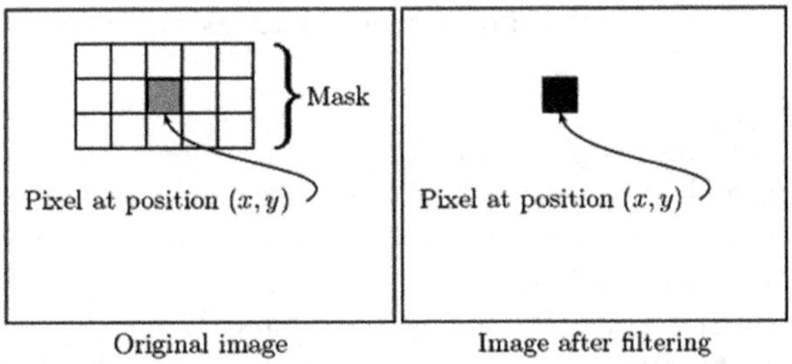

Fig. 4.6 Neighborhood processing

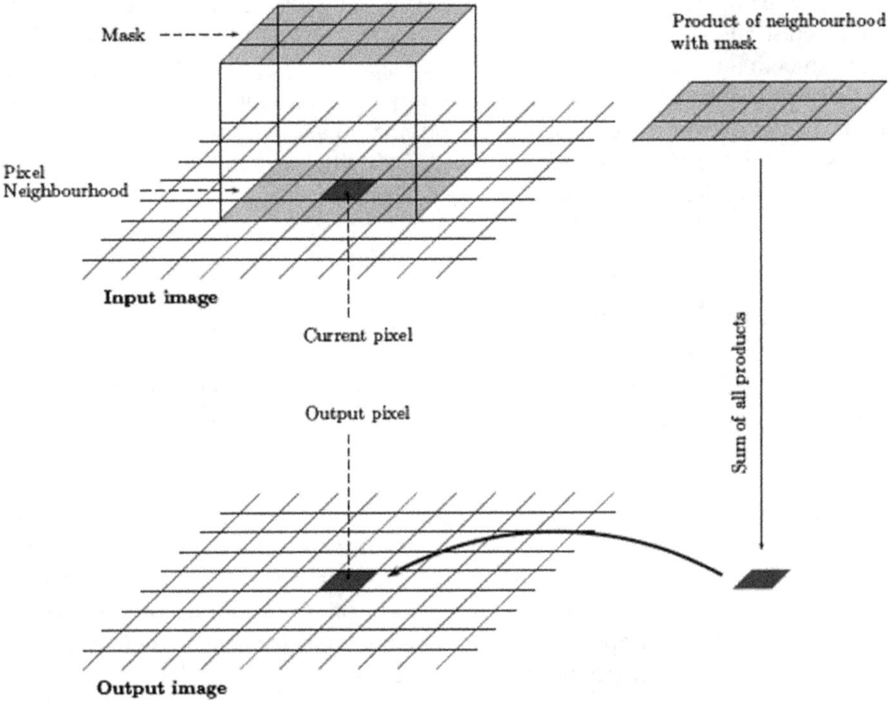

Fig. 4.7 Image filtering in spatial domain

An image filter combines a mask with an operating function. This filter is called a linear filter if it produces a new gray-level value using linear operations. The filter can be implemented by multiplying all values in the mask by corresponding values in neighboring pixels and adding them together. Let us consider a 3 × 5 mask, as shown in Fig. 4.7. This process is known as spatial filtering. It is used in the convolution layer of a convolutional neural network (CNN).

4.3.1.2 Image Enhancement in Transform Domain

Gray-level values change with distance, creating a frequency. A high-frequency coefficient is defined by large changes in gray levels over a short distance, for example, noise and edges. Small changes in gray-level values over large distances define low-frequency coefficients, i.e., background. Based on these frequencies, low-pass and high-pass spatial filters are used for image enhancement in the spatial domain. High-pass filters pass high-frequency coefficients while blocking low-frequency coefficients. Backgrounds and skin textures are removed using this filter. Low-pass filters pass low-frequency coefficients and eliminate high-frequency coefficients. This filter removes noise and edges from the image. The low-pass filter is the smoothing filter, and the high-pass filter is the sharpening filter. The Fourier domain is used to enhance images in the frequency domain. An image is filtered in the spatial domain by convolving the original image with a filter mask. Signal processing fundamentals say a frequency-domain filter process would be an image multiplied by a filter transfer function to get the filtered image.

4.3.2 Image Restoration

Images obtained through the acquisition process are not the exact same information as represented objects in the image, but there is some degradation in the acquired images. During the acquisition process of an image, many sensors or devices can cause degradation. For example, in remote sensing and astronomy, images are degraded due to various atmospheric conditions, various lighting conditions in space, and the camera position of satellites. In many applications, point degradations (due to noise) and spatial degradations (due to blurring) are commonly used to degrade images. Restoration of an image is restoring an original image from a degraded one. Rotation appears similar to image enhancement by definition, but there are some differences between the two processes.

- The image enhancement process is subjective, while the image restoration process is objective.
- Image enhancement procedures utilize the psychophysical aspects of the human visual system (HVS) to manipulate an image. To reconstruct the original image, images are restored by modeling degradation and applying inverse processes.
- Quantitative parameters cannot be used to measure image enhancement. Quantitative parameters can be used to measure image restoration.
- Contrast stretching is an example of image enhancement. Removing blur from an image is an example of image restoration.

4.3.3 Image Morphology

Various image processing operations that deal with the shape of features in an image are known as morphological image processing (or morphology) [1, 5]. A binary image can be corrected with these operations by correcting the image's imperfections. Variously shaped structuring elements can extract shape features (such as edges, holes, corners, and cracks) from an image. This process is used in industrial computer vision applications, such as object recognition, image segmentation, and defect detection. This process involves various operations, such as erosion, dilation, opening, closing, etc., used to process images.

4.3.4 Image Segmentation

Segmentation involves splitting an image into constituent parts based on some image attributes. This process reduces excessive data while only important data is retained for image analysis. Additionally, this process converts bitmap data into more readable structured data. Segmenting images using the similarity and discontinuity properties of pixels is possible. A similarity property in pixels means that targeted pixels have the same gray-level intensity. In contrast, a discontinuity property means that boundary pixels have different gray levels in targeted pixel groups. During the image segmentation process, three features are extracted from the image: lines, points, and edges. In a similar way to mask processing, detection features are detected. The operations such as line detection, point detection, and edge detection are associated with the image segmentation process. The various edge detectors, such as Prewitt, Sobel, Roberts, and Canny, are used in real-world applications according to their requirement in the use case.

4.3.5 Image Compression

A digital image compression process reduces the amount of redundant and irrelevant information in the image data to store or transmit it efficiently. Redundancy in the image can be classified into coding redundancy, interpixel redundancy, and psychovisual redundancy.

- **Coding Redundancy:** In images, a few bits are used to represent frequently occurring information. An image is represented by its pixel values. It is called a code when these symbols are used. Each pixel value in an image is assigned a code word. Usually, look-up tables (LUTs) are used to implement this type of code. Huffman codes and arithmetic coding are examples of image compression methods that explore coding redundancy.

- **Interpixel Redundancy:** It refers to the correlation between pixels next to each other in the image. The term spatial redundancy is also used to describe it. Here, the information carried by an individual pixel is almost related to its nearest pixels. Run-length encoding (RLE) and many predictive coding methods, such as differential pulse code modulation (DPCM), explore this redundancy.
- **Psychovisual Redundancy:** As a result of the human visual system's (HVS's) ignoring of data, this redundancy occurs. Psychovisual redundancy is reduced using the quantization process. JPEG encoding standard explores this redundancy through image compression.

There are two types of image compression: lossy and lossless. A lossless compression method is preferred for archival purposes and is often used for medical imaging, technical drawings, clip art, or comic books. It is common for lossy compression methods to introduce compression artifacts, mainly when used at low bit rates. In applications where a small loss of fidelity (sometimes invisible) is acceptable to reduce bit rates substantially, lossy methods are ideal for native images like photographs. In some cases, visually lossless compression may be called lossy compression which produces negligible differences. Compressing data with lossy methods such as transform coding, discrete cosine transform, color quantization, chroma subsampling, and fractal compression is possible. Several lossless compression methods exist, including run-length coding, area image compression, predictive coding, entropy coding, and adaptive dictionary algorithms like Lempel-Ziv-Welch (LZW), DEFLATE, and chain coding.

4.3.6 Image Registration

Registration of images involves transforming different sets of data into one coordinate system. The data may be multiple photographs, data from various sensors, data from multiple depths, or data from different viewpoints. The technology is used in computer vision, medical imaging, automatic target recognition in the military, and compiling and analyzing satellite images. This data must be registered to be compared or integrated. Image can be registered using various methods such as point matching, feature matching (e.g., scale-invariant feature transform (SIFT)), etc.

4.3.7 Object Detection

Object detection identifies objects in an image and is commonly used in surveillance and security applications. Nowadays, convolutional neural networks (CNNs) are widely used for object detection, but other algorithms like region-based CNN (R-CNN), fast R-CNN, and you only look once (YOLO) can also be implemented for this purpose.

4.3.8 Image Manipulation

Image manipulation is the process of altering an image to change its appearance. There may be several reasons for this, such as removing unwanted objects from images or adding objects not present in the image. Graphic designers use this process to create posters, films, etc.

4.4 Industrial Applications of Image Processing

Many industries rely on digital image processing, including advertising, marketing, design, photography, etc. In the medical field, digital image processing is used for many applications, such as X-ray imaging, CT scans, etc. With satellites, remote sensing scans the Earth and records its features. Machine vision or robot vision is an application of digital image processing that uses the software. It takes a lot of time and effort to process digital images, but it will result in a higher quality of life for humans.

A wide variety of industries can benefit from image recognition technology. Several enterprises have adopted this technology due to better manufacturing, inspection, and quality assurance tools and processes, making them more time-efficient and productive. Large corporations and startups, such as Google, Adobe Systems, etc., rely heavily on image processing techniques daily. Over the next few years, AI (artificial intelligence) will advance this technology significantly. Here, information regarding a few industries is given where image processing significantly improves some operations.

4.4.1 Agriculture

A large amount of image processing is also used in agriculture. This technology offers the advantage of being nondestructive, providing insight into crops without touching them. Irrigation and pest control are two of the first aspects of crop management. This can be achieved by various image processing with the help of machine learning.

Nowadays, image processing is commonly used to identify diseased plants in agriculture. Traditionally, an expert would be consulted for this task. The good news is that image processing technology can save the day in this case. In the preprocessing phase, the digital images are improved in terms of resolution, noise, and color. A database will be used to refer the enhanced image to related images once it has been segmented. Afterward, the segmented image is compared to a reference image to determine if it contains defects.

4.4.2 Manufacturing

Digital image processing has many possible industrial applications; therefore, many industries are interested in it. Quality assurance in manufacturing, outgoing inspections, and other areas are some of the main applications of industrial quality assurance. The technologies cover a broad spectrum, from physical inspection of surfaces to 3D optical measurements, X-rays, heat flow thermography, terahertz measurements, and nondestructive testing [6]. The various industrial areas where image processing can be used and pointed by the Fraunhofer Research Institute, Germany [6], such as:

- Detection of defects in various industrial parts such as complex structures, contamination, and production line components
- Quality inspection of product and other important components related to production such as carbon fibers, belt material, heat flow thermography, etc.
- Sorting of various materials used in manufacturing
- Visualization of various structures related to production and measure distance and position of different components during production

4.4.3 Automotive

Almost every automotive system has a camera and image processing system [7]. High-speed systems must detect micrometer variations from the target value to achieve 100% quality control on the production line. Intelligent imaging systems also provide insights into other fields, such as automobile driving, traffic control, crash laboratories, and wind tunnels. With robust housings and electronics, high-speed cameras can capture all possible angles inside and outside the crashed vehicles, capturing the scene from every possible angle. The HD quality of the images ensures that engineers can follow every detail of the deformation of car bodies. Fast-moving manufacturing processes require high-speed cameras to analyze faults in detail. A camera's superiority is clear when it comes to high-speed processes. Imaging systems are increasingly taking over quality and process control in mainstream processing. Image processing ensures brilliant surfaces around the clock, micrometer-accurate assembly tolerances, and defect-free circuits on increasingly prevalent chips and microcontrollers.

The cost of errors is synonymous with the cost of production for automotive manufacturers. Automotive materials are produced with glass-like transparency using cameras and laser systems. Engine developers are gaining a deeper understanding of the processes behind injection and combustion that are not visible to the naked eye. Camera systems can see and analyze even the slightest turbulence in wind tunnels. Colleagues use laser systems for interior design, tire development, and vehicle body design to detect and assess vibrations and structure-borne sounds. It is not only used for diagnosis but also for measuring and documenting the effectiveness of their measures.

Fully automated processes become more flexible with robotics. Image processing software calculates the location and plans based on the images captured by cameras. A sensor cluster containing several cameras allows highly accurate 3D coordinates to be determined for large objects. The Six Sigma approach to quality management in the automotive industry matches 100% real-time inspection on the production line. According to the control loop define-measure-analyze-improve-control, manufacturers and major suppliers strive to achieve a zero defect objective. This is made possible by camera systems combined with downstream analysis software.

4.4.4 Healthcare

Image processing in healthcare is mainly called medical image processing or analysis. Nowadays, applications and research are developing in the healthcare sector using different kinds of medical images. Analyzing medical images is often done using computational methods to extract meaningful information. The task of medical image analysis involves visualizing and exploring 2D images and 3D volumes and segmenting, classifying, registering, and reconstructing 3D images. A variety of imaging modalities may be used for this analysis, including X-ray (2D and 3D), ultrasound, computed tomography (CT), magnetic resonance imaging (MRI), and nuclear imaging (PET and SPECT).

4.4.5 Robotics Guidance and Control

A robot uses images for certain robotic tasks. Imagery equipment and the necessary programming and software can be available from robotics specialists to handle visual input encountered by robots. Robots are taught to recognize and respond to images as part of the programming and teaching process. Software suites are available from some companies for direct installation on equipment, or you may program your own. In robotics, a camera system is used for navigation as an example of image processing. There are many ways to teach robots to follow lines, dots, or other visual cues, such as lasers. Targets in the surrounding environment are identified and tracked using a crude camera and image processing system. In a factory, this can be helpful for automating processes like collecting and delivering products by robots.

4.4.6 Defense and Security

Several digital image processing applications are widely used in defense and security, including small target detection and tracking, missile guidance, vehicle navigation, wide-area surveillance, and automatic/aided target recognition [8]. In defense and security applications, image processing can reduce the workload of human analysts so that more image data can be collected in an ever-increasing volume. Researchers who work on image processing also aim to develop algorithms and approaches to facilitate autonomous systems' development. This will enable them to make decisions and take action based on input from all sensors.

References

1. R.C. Gonzalez, R.E. Woods, *Digital Image Processing* (Pearson Education India, Upper Saddle River, 2008)
2. The University of South Carolina SIPI Image Database. http://sipi.usc.edu/database/database.php. Last Access January 2023
3. R. Kundu, *Image Processing: Techniques, Types, & Applications* (2023). Weblink: https://www.v7labs.com/blog/image-processing-guide. Last Access Jan 2023
4. R.C. Gonzalez, R.E. Woods, *Digital Image Processing Using MATLAB* (TATA McGraw-Hill Education, New York, 2009)
5. N. Efford, *Digital Image Processing: A Practical Introduction Using JAVA* (Pearson Education, London, 2000)
6. Application of Industrial Image Processing. Web link: https://www.vision.fraunhofer.de/en/application-of-industrial-image-processing.html. Last Access Jan 2023
7. Image Processing in the Automotive Industry (2019). Web link: https://www.industr.com/en/image-processing-in-the-automotive-industry-2356834. Last Access Jan 2023
8. E. Du, R. Ives, A. van Nevel, J.H. She, Advanced image processing for defense and security applications. EURASIP J. Adv. Signal Proc. **2010**(1), 1–1 (2011)

Chapter 5
Artificial Intelligence and Its Applications

Artificial intelligence is a computer system that can accomplish tasks normally handled by humans. Machine learning and deep learning are used to power these systems. The field of machine learning (ML) falls under the umbrella of artificial intelligence (AI) [1]. Machine learning algorithms enable systems to learn automatically. This system allows learners to improve their learning experience without learning complex programming techniques. A precious aspect of machine learning is developing a new model based on computer systems and programs that access information and use it to learn for themselves [1]. These algorithms determine unique features or patterns in the input data to make better decisions. Several applications use these algorithms, including medical image processing, computer vision, recognition of biometrics, object detection, and automation, among others. Supervised, unsupervised, and reinforcement learning are all types of machine learning [1, 2].

In real-time applications related to machine learning, various kinds of data, such as text, images, videos, speech signals, etc., are used as input data. The basic steps of the machine learning algorithm are given in Fig. 5.1. Training and testing are two phases of the machine learning algorithm. First, the model learns unique features or patterns from the input image during the training phase. Then, the model produces specific outputs based on learner features or patterns in the testing phase. For example, an image may have features such as edges, a region of interest, etc., which can be extracted using various feature extraction methods. The selection of extraction methods depends on the type of input data and the specific output of the model.

R. Thanki, P. Joshi, *Advanced Technologies for Industrial Applications*,
https://doi.org/10.1007/978-3-031-33238-8_5

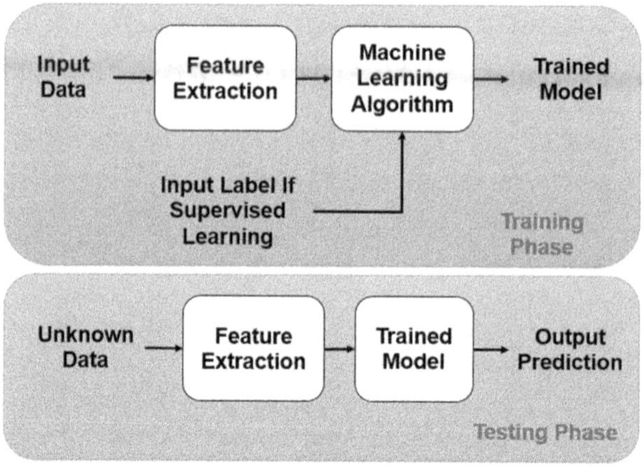

Fig. 5.1 Basic operation of machine learning

5.1 Types of Learning Methods

The details of different learning methods are covered in this section.

5.1.1 Supervised Learning

This type of learning is often used using real-time applications and practical approaches. The model learns information by analyzing the previous experiences it has had with the information provided. This type of learning involves mapping an input (x) to an output (y) by an algorithm that gives a mapping function (f) like this:

$$y = f(x) \tag{5.1}$$

Classification and regression are two types of supervised learning related to problems. There are three classification problems: value, group, and category. As an example, classification of "cat" or "dog". The regression problem involves a continuous or real value as the output, like temperature or currency. Various methodologies are used to predict the output of supervised learning algorithms [3].

5.1.2 Unsupervised Learning

An algorithm that learns by itself attempts to discover a unique pattern or feature without prior knowledge of the pattern or feature. A mathematical model with input (x) but no corresponding output is considered this type of learning. An unsupervised learning system finds its answer to input but does not provide the correct answer. Association and clustering problems are typically solved using algorithms based on unsupervised learning.

5.1.3 Reinforcement Learning

A machine or system learns by taking a particular action to maximize output given a given input. Various software and algorithms determine the machine's optimal possible outcome or behavior. Unfortunately, machine learning algorithms are not suitable for all real-time applications. The task of finding practical algorithms is, therefore, a trial-and-error one. To solve this problem, researchers [1–5] recommend that the algorithm choice depends on the input data type and size. As a result, machines are used for reinforcement learning in practice. In this type of learning, the algorithm is continuously exposed to it and trained to predict output better.

5.1.4 Deep Learning

Deep learning is a subset of artificial intelligence and a newly developed learning method that uses neural networks [6] for output prediction. Nowadays, researchers in the literature propose many models based on this learning. In 1971, Dr. Robert Hecht-Nielsen, the inventor of the first neurocomputer, introduced the first neural network (NN) [6]. Artificial neural networks (ANN) are basically these networks. In his definition, a neural network is a computer system comprised of many simple, highly interconnected processing elements whose dynamic states respond to external inputs [6]. Application areas for this network include Big Data analysis, person recognition, and data prediction. The forwarding neural network (FNN) is also called this network. The network uses neurons arranged in tiers based on parallel operations. Figure 5.2 shows a simple artificial neural network. Input, hidden, and output layers are the main layers of the network. A neuron's or node's number depends on the size of its inputs and outputs. Nodes are fully connected to each other. Links with specific weighting values connect two nodes of adjacent layers.

5.2 Types of Machine Learning Algorithms

The details of different machine learning algorithms are covered in this section.

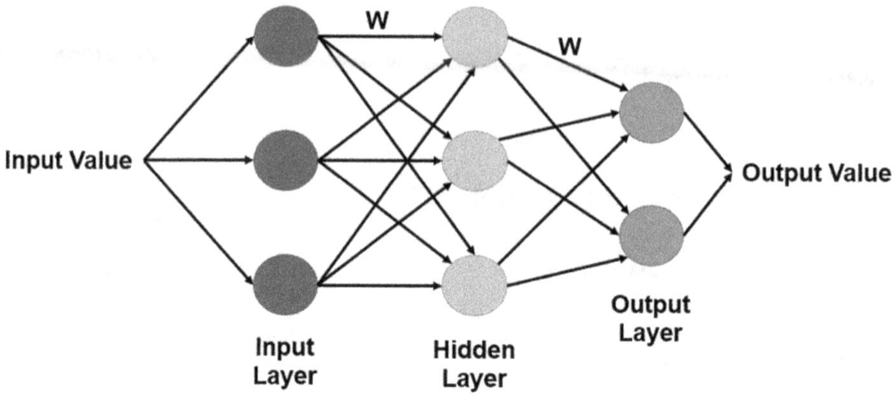

Fig. 5.2 Basic structure of artificial neural network

5.2.1 Supervised Learning-Based Algorithms

For the testing of this algorithm, prior knowledge of the dataset is essentially necessary. It is the analyst's responsibility to gather this dataset knowledge. Here are the steps in this algorithm [7]:

- For each input data class, identify the training areas.
- It identifies the data's mean, variance, and covariance.
- The data is then classified.
- Finally, the input class has been mapped.

These algorithms have the advantage of detecting and correcting errors during evaluation. Time-consuming and costly are the main disadvantages of these algorithms. Additionally, the researcher, scientist, or analyst may not consider all conditions affecting the dataset's quality when selecting a training dataset. This led to human error in the performance of these algorithms.

5.2.1.1 Statistical Learning-Based Algorithms

Based on mathematical theories, statistical learning-based classifiers aim to predict some meaningful output by finding relationships between classes. On smaller datasets with fewer attributes, these classifiers are applied. There are several statistical learning-based classifiers, such as minimum distance (MD), Mahalanobis distance (MhD), and maximum likelihood classifiers (MXL), available in the literature [8]. These classifiers are discussed in detail in Lillesand and Keifer [9]. Based on Bayesian probability theory, MXL is widely used in image classification. This algorithm uses a matrix of Gaussian distributed dataset patterns and its covariance matrix to calculate the probability of the input dataset.

5.2.1.2 Nearest Neighbor (NN) Algorithm

A famous machine learning algorithm for data classification is nearest neighbors. For example, it classified the types of images based on the input dataset of their nearest neighbors in the image dataset. It predicts that objects near each other have similar characteristics. It is a nonparametric algorithm that does not require any assumptions about the distribution of the input dataset. For the identification of relevant attributes, some prior knowledge of the input dataset is required.

5.2.1.3 Naive Bayes Algorithm

An algorithm for supervised machine learning based on Bayes' theorem and the "naive" assumption of independent features from each training and test dataset is presented [10].

5.2.1.4 Support Vector Machine (SVM) Algorithm

Vapnik proposed the support vector machine (SVM) in 1995 [11]. A boundary decision (hyperplane) is used in this classifier to separate input data belonging to one class from input data belonging to another class. With linear functions separating input data from output data, an SVM's optimized hyperplane has the largest margin. The loss function is used if the input data is separated by a nonlinear function. Not linearly separated data is transformed into linearly separated data by SVMs using different kernel transforms. An SVM commonly uses three kernel functions: polynomial learning machines, radial-based function networks (RBFN), and two-layer perceptions. RBFN is generally used for training classifiers because it is more powerful and effective than the other two kernel functions [11, 12]. A classifier like this can effectively classify input data into two classes but can also classify data into multiple classes using error-correcting output codes. It is very easy to understand and has been proven to be accurate.

5.2.1.5 Decision Tree Algorithm

The decision tree algorithm can solve regression and classification problems [13]. Learning decision rules from the training dataset creates a model that can classify classes. This algorithm is very simple to understand compared to other supervised learning algorithms. A tree structure representation is used in this algorithm to solve the problem. Tree nodes represent dataset attributes, and leaf nodes represent class labels. A decision tree is a classifier capable of classifying multiclass input datasets. Several decision tree algorithms are available in the literature [14], such as ID3, C4.5, C5.0, and CART. Ross Quinlan developed ID3 in 1986, also known as Iterative Dichotomiser 3. There are multiple trees for each categorical feature of the data given by the algorithm. A C4.5 algorithm replaces ID3 and converts trained trees into if-then rules. The C5.0 algorithm is the latest version of the ID3 algorithm. Classification and regression trees, called CART, are similar to the C4.5 algorithm. However, they support numerical target variables and do not require computing sets to construct trees.

5.2.1.6 Random Forest Algorithm

It is based on constructing multiple decision trees using the random forest algorithm [15–17]. The input value for each class should be placed on each tree of the forest when classifying a novel class from an input dataset using a random forest algorithm. An average value is calculated and assigned a new classification based on each tree's classification. A random forest algorithm consists of two stages: creating the random forest and predicting the classifier based on the generated random forest.

5.2.1.7 Linear Regression Algorithm

A linear regression model analyzes the relationship between an independent variable (x) and a dependent variable (y) by modeling their linear relationship. Linear refers to a straight line between independent and dependent variables. Other things are to be kept in mind. As x increases/decreases, y also changes linearly. Mathematically, the relationship can be expressed as follows:

$$y = Ax + B \tag{5.2}$$

In Eq. 5.2, A and B are the constant factors. In the supervised learning process using linear regression, the goal is to find the exact value of constants "A" and "B" using the datasets. Using these values, i.e., the constants, you can predict the future value of "y" for any value of "x." Specifically, a simple linear regression involves a single independent variable, whereas multiple linear regression is used if there is more than one independent variable.

5.2.1.8 Logistic Regression Algorithm

The logistic regression algorithm is often used for classification in supervised machine learning. The name "regression" can be misleading, so let's not consider it a regression algorithm. Logistic regression takes its name from a special function called the logistic function, which plays a central role in the process. The logistic regression model can be described as a probabilistic model. The probabilities of an instance belonging to a certain class can be determined using this method. The output is between 0 and 1 since it is probabilistic. We can consider positive and negative classes whenever we use logistic regression as a binary classifier (classification into two categories). The probability is then calculated. If the probability is greater than 0.5, it is more likely to fall into the positive category. We can also classify this as negative if the probability is low (less than 0.5).

5.2.2 Unsupervised Learning-Based Algorithms

Clustering algorithms are also known as unsupervised learning algorithms. In contrast to supervised methods, these algorithms require a minimum amount of input data to be analyzed. Using these algorithms, the data is grouped into groups containing the same information. Instead of categorizing the training user data, the system selects the mean and covariance of the class. This algorithm is called unsupervised classification because the classification process depends on the system. A user can define how many classes or clusters to create. Each cluster can then be assigned key information for easy analysis after classification. Researchers have developed many clustering algorithms in terms of accuracy and decision-making rules. To achieve optimal output in these algorithms, iterative calculation of input data is used.

It is possible to perform these algorithms in two steps [18]:

- Identify possible clusters within a dataset or image in the first step.
- Using the distance measure, assign each pixel a cluster based on the distance between the data or on an individual pixel basis [18].

The general steps for this algorithm are as follows [18]:

- This algorithm requires the following information: radius for cluster area, merging parameters for the cluster, and number of pixels evaluated. Cluster identification is the process of identifying groups of clusters within a data or an image.
- Various labeling is assigned to clusters within a data or image for proper analysis.

5.2.2.1 K-means Clustering Algorithm

The K-means algorithm [18–26] is a well-known method for predicting unsupervised data. This algorithm classifies all pixels based on their distance from the cluster mean [1].

Once classification is done, the updated mean vectors for each cluster are calculated. This process will be performed for a number of iterations until there is no variation in the location of cluster mean vectors between two successive iterations [18]. The main objective of this algorithm is to estimate variation within a cluster. The K-means algorithm performs two steps: the location of initial cluster centers and subsequent cluster merging.

5.2.2.2 Principal Component Analysis

Principal component analysis (PCA) [27], also known as Karhunen-Loeve analysis, transforms the data into new transforms that better interpret the original data. It compresses the data information into a few principal components of data. The description of PCA is beautifully described by Schowengerdt [28] and Gonzales-Wood [29].

5.2.2.3 Independent Component Analysis

Clustering can be improved using independent component analysis (ICA) [30–32]. This is accomplished by assuming non-Gaussian pixel values and statistical independence between the subcomponents. An ICA method is used to segment images blindly.

5.2.2.4 Singular Value Decomposition

Many unsupervised classification approaches use singular value decomposition (SVD) [33–35]. The SVD method is widely used in data clustering as a preprocessing and dimensionality reduction method. SVD will reduce the data and then classify it [33].

5.2.2.5 Gaussian Mixture Models

Gaussian mixture models (GMMs) [36–40] can be used to cluster, recognize patterns, and estimate multivariate density [39, 40]. This algorithm offers the advantages of being easy to implement, being computationally fast, and giving tighter clusters.

5.2.2.6 Self-Organizing Maps

Using self-organizing maps (SOMs) [41–47], data can be visualized on hexagonal or rectangular grids. These tools are used in various fields, including meteorology, oceanography, project prioritization, and oil and gas exploration. Kohonen maps [41] and self-organizing feature maps (SOFMs) are also self-organizing maps. Neurons are arranged in a multidimensional network in this algorithm.

5.2.3 *Reinforcement Learning-Based Algorithms*

Reinforcement learning (RL) algorithms [48–50] are machine learning algorithms that allow machines to automatically determine data's behavior in a specific manner to increase its performance. These algorithms have the main limitation in that they require some learning agent. A reinforcement learning algorithm is designed to solve a specific type of problem. Depending on its current state, an agent is supposed to decide what is the appropriate solution. A Markov decision process occurs when this process is repeated. Many reinforcement learning algorithms are available in the literature [48], including Q-learning, temporal difference, and deep adversarial networks. The following are the steps that should be followed for these algorithms:

- The agent observes the input state.
- Agents perform actions based on their decision-making functions.
- The agent receives a reward or reinforcement after the action has been performed.
- Information about the state-action pair is stored.

5.2.3.1 Basic of RL Algorithm and Q-Learning Algorithm

A learning agent and an environment are the two components of any RL algorithm. Agents refer to the RL algorithm, while environments refer to objects they act on. Initially, the environment sends a state to the agent, responding to it based on its knowledge. The environment sends the agent a pair of next-state values and rewards in the next step. The agent uses a reward returned by the environment to evaluate its last action by updating its knowledge. Loops continue until an episode is terminated by the environment. Q-learning algorithms is an off-policy, model-free RL algorithm.

5.2.3.2 State-Action-Reward-State-Action (SARSA) Algorithm

There are many similarities between SARSA and Q-learning. SARSA is an on-policy algorithm, whereas Q-learning is not. Instead of learning the Q-value based on greedy policy actions, SARSA learns it based on current policy actions.

5.2.3.3 Deep Q Network (DQN) Algorithm

The main weakness of Q-learning is its lack of generality, although it is a very powerful algorithm. A dynamic program is similar to Q-learning because it operates on numbers in a two-dimensional array (action space + state space). Q-learning has no idea what action to take for states it has never seen before. Hence, Q-learning agents can't estimate values for unseen states. DQN introduces neural networks as a solution to this problem. The DQN estimates the Q-value function using a neural network. The network outputs the corresponding Q-value for each action based on the input current.

5.3 Types of Deep Learning Algorithms

The details of different deep learning algorithms are covered in this section.

5.3.1 Convolutional Neural Networks (CNNs)

Convolutional neural networks (CNNs) are the most commonly used deep learning neural networks for image-related applications [51–54]. The CNN has three layers: an input, an output, and many hidden layers. Figure 5.3 shows CNN's basic architecture. Several operations are carried out in the hidden layer of CNN, such as feature extraction, flattening of features, and classification of features.

5.3.1.1 Feature Extraction Operation

As part of the feature extraction operation, features are convolutioned, rectified linear units (ReLUs) are constructed for nonlinearity, and pools are created. The following is a description of how each task is performed:

- **Convolution:** In CNN, the first step is convolution, which is the process of extracting features from an input image. The process is similar to the spatial filtering of an image, using small information from an input image to determine

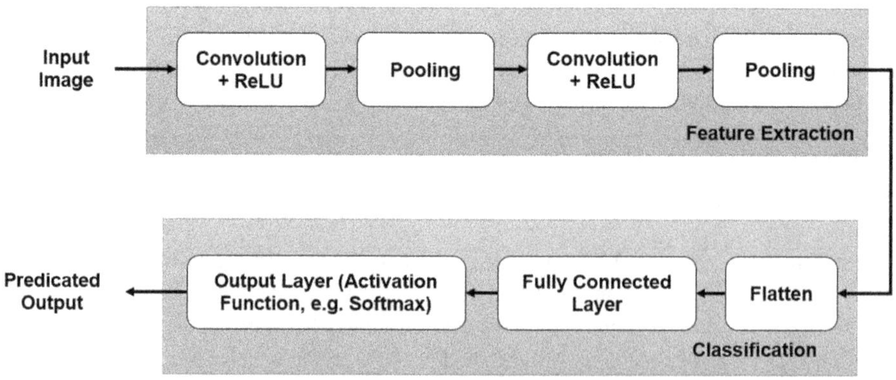

Fig. 5.3 Basic architecture of CNN

the relationship between pixels. Based on math, it is an output that uses two input values: the value of the image pixel and the value of the filter mask. Strides and padding are also used to extract the features more effectively after the convolution process. The operation of the steps is used to get better features from input images. Padding may be necessary when a filter is not applied perfectly to an input image. The filter works effectively on images with zero values in this operation.

- **Nonlinearity ReLU:** Nonlinearity ReLUs are rectified linear units with nonlinear operations performed on convolved features. In essence, it removes negative values from the convolved features. It can perform various operations, such as maximum, minimum, mean, etc.
- **Pooling:** Pooling operation such as upsampling or downsampling reduces the dimensions of each feature by reducing the dimension size. To reduce the dimensions of extracted features, CNN uses pooling operations such as max, sum, and average.

5.3.1.2 Classification Operation

Three different operations are involved in this operation: flattening, prediction of features, and activation. In flattening, features are extracted from the input image and converted to vectors. For the prediction of the input feature vector, a fully connected network, such as a neural network, is fed this vector. As a final step, the predicted output of the neural network is classified using an activation function like softmax or sigmoid.

5.3.2 Other Deep Learning Algorithms

Image-related applications are investigated using a variety of deep learning (DL) algorithms [6]. Among these algorithms are convolutional neural networks (CNNs), deep autoencoders (DA), recurrent neural networks (RNN), deep belief networks (DBN), and deep neural networks (DNN), along with deep conventional extreme machine learning (DC-EML) techniques available in the literature [6]. The advantage and disadvantages of these models are given in Table 5.1. The various types of CNN architectures, such as AlexNet [55], LeNet [56], faster R-CNN [57], GoogLeNet [58], ResNet [59], UNet [60], etc., are available for different types of real-time applications.

5.4 AI-Based Research in Various Domains

In the last 10 years, research and development have been conducted in AI-based systems within various areas, such as developing new learning algorithms

Table 5.1 Advantage and disadvantage of deep learning algorithms

Sr. no.	Deep learning model	Advantages	Disadvantages
1	DNN	Widely used	Required more training time
2	CNN	Fast learning process	Required Big Data
3	RNN	Used in sequential	Required Big Data
4	DBN	Greedy norm used in prediction	Complex algorithm
5	DA	No labeled data required	Required pre-training process
6	DBM	More robust against interference	Not used for Big Dataset

and models, computer vision, natural language processing, robotics, recommender systems, the Internet of Things, and Advanced Game Theory. The details of each research domain where AI is used nowadays are given below [61, 62]:

5.4.1 Development of New Algorithms and Models

Various algorithms or models are being developed for machine learning and deep learning everywhere. Due to the development of new models, more data availability, and fast computing capabilities, AI applications have increased exponentially. Healthcare, education, banking, manufacturing, and many other industries have AI applications. A major challenge in AI-based projects is improving model performance. A single structured process cannot guarantee success when implementing ML and DL applications in business at this time. A model's performance is one of the most important factors in developing an AI model. It is mainly a technical factor that determines model performance. Deploying a machine learning or deep learning model that isn't accurate enough for the output makes no sense for many use cases.

5.4.2 AI in Computer Vision

In computer vision, computers and systems can detect and interpret meaningful information from digital images, videos, and other visual signals and then take appropriate actions or recommend further actions. The following are a few research areas of well-known computer vision tasks [63]:

- **Image Classification:** Seeing an image, image classification can classify it (a dog, an apple, a face). The algorithm is capable of accurately predicting what class an image belongs to. It might be useful to a social media company to automatically identify and filter objectionable images uploaded by users.
- **Object Detection:** Images and videos can be classified using image classification to identify a certain type of image, which can then be detected and tabulated. A

typical example would be detecting damage on an assembly line or identifying machinery that needs maintenance.

- **Object Tracking:** An object is tracked once it has been detected by object tracking. The task is frequently performed by capturing images in sequence or viewing live video feeds. Autonomous vehicles must classify and detect pedestrians, other cars, and road infrastructure and track them in motion to avoid collisions and adhere to traffic laws.
- **Content-Based Image Retrieval:** With content-based image retrieval, images are browsed, searched, and retrieved from large data stores based on their content rather than their metadata tags. A technique that replaces manual image tagging with automatic annotation can be incorporated into this task. Digital asset management systems can use these tasks to improve search and retrieval accuracy.

5.4.3 AI in Natural Language Processing

Artificial intelligence that uses computer software to understand text and speech input in the form of natural language is known as natural language understanding (NLU). In NLU, humans and computers can interact with each other. Computers can understand commands without the formal syntax of computer languages by comprehending human languages, such as Gujarati, Hindi, and English. In addition, NLU allows computers to communicate with humans in their own language. Two main techniques are used in natural language processing: syntax and semantic analysis. The syntax of a sentence determines how words are arranged to make grammatical sense. Language processing uses syntax to determine meaning from grammatical rules within a language. The main research areas in AI-based NLU are text processing, speech recognition, and speech synthesis.

5.4.4 AI in Recommender Systems

Several information filtering systems, including recommender systems (sometimes called recommendation engines or platforms), provide users with suggestions for items relevant to their needs. In most cases, the suggestions relate to how to make decisions, such as buying something, listening to music, or reading online news. A recommendation system can be beneficial when a person chooses an item from many options a service offers. Video and music services use recommender systems to create playlists, online stores use product recommenders, and social media platforms use content recommenders to recommend content. There are a variety of recommender systems available. These systems can operate with a single input, such as music, or multiple inputs, such as news, books, and search queries, within and across platforms. Restaurants and online dating sites also have popular recommendation systems. Recommendations and systems have also been developed to assist researchers in searching for research articles, experts, collaborators, and financial services. A few hot research topics for AI in recommender systems are [64] the development of algorithm-based deep learning and reinforcement learning, the development of knowledge graph-based algorithms, and explainable recommender systems.

5.4.5 AI in Robotics

Most problems associated with robotic navigation have been solved, at least when working in static environments. Currently, efforts are being made to train a robot to interact with the environment in a predictable and generalizable manner. Among the topics of current interest is manipulation, a natural requirement in interactive environments. Due to the difficulty of acquiring large labeled datasets, deep learning is only beginning to influence robotics. Reinforcement learning, which can be implemented without labeled data, could bridge this gap. However, systems must be capable of exploring policy spaces without harming themselves or others. A key enabler of developing robot capabilities will be advanced in reliable machine perception, including computer vision, force, and tactile perception. These advances will be driven in part by machine learning.

5.4.6 AI in the Internet of Things

Sensory information can be collected and shared by interconnecting various devices. Research into this idea is growing. A device in this category would include an appliance, vehicle, building, camera, etc. In addition to using wireless networking and technology to connect the devices, AI can utilize the data produced for intelligent and useful purposes. The current array of communication protocols used by these devices is bewildering. This research problem might be solved by artificial intelligence.

5.4.7 AI in Advanced Game Theory

A growing body of research examines artificial intelligence's economic and social dimensions, including incentive structures. Many academic institutions have studied distributed AI and multi-agent systems since the 1980s, and their popularity increased following the Internet's arrival in the late 1990s. Ideally, systems must be able to handle conflicts of interest among participants or companies, such as self-interested humans or automated AI agents. Various topics are being explored, including computational mechanism design, computational social choice, and incentive-aligned information elicitation.

5.4.8 AI in Collaborative Systems

A collaborative system is an autonomous system that can collaborate with other systems and humans using models and algorithms. Formal collaboration models are developed in this research, and the capabilities needed for systems to be effective partners are examined. Humans and machines can work together to overcome the limitations of AI systems, and agents can augment human abilities and activities. As a result, diverse applications have emerged that utilize the complementary strengths of humans and machines.

5.5 Industrial Applications of AI

AI can be used for anything from solutions relevant to consumer-friendly to highly complex industrial applications, such as predicting the need for manufacturing equipment maintenance. Examples from several industries illustrate the breadth and depth of AI's potential [65, 66].

5.5.1 Financial Applications

In the consumer finance area as well as in the global banking sector, artificial intelligence has many applications. In this industry, artificial intelligence can be found in the following applications:

1. Fraud Detection: In recent years, financial fraud has been committed on a massive scale and daily. These crimes cause major disruptions for individuals and organizations.
2. Stock Market Trading: Stock market floor shouting is a thing of the past. Most major trading transactions on the stock markets are handled by algorithms that make decisions and react much faster than humans ever could.

A unique application of artificial intelligence can be found in the insurance world within the broader financial services landscape. A few examples are:

1. AI-Powered Underwriting: There have been a lot of manual processes used to make underwriting decisions for decades, and data inputs like medical exams have been added to the mix. As a result of artificial intelligence, insurance com-

panies use massive datasets to assess risks based on factors such as prescription drug history and pet ownership.

2. Claims Processing: Artificial intelligence can handle simple claims today. A simple example of it is chatbots. Human involvement in claims decisions will likely decrease as machine vision and artificial intelligence capabilities increase.

5.5.2 Manufacturing Applications

AI adoption is highest in the industrial manufacturing industry, with 93% leaders claiming their companies use it moderately or more. Manufacturers' most common challenges are equipment failure or delivery of defective goods. As manufacturers take steps toward digital transformation, AI and machine learning can improve operational efficiency, lunch new or updated products, customize product designs, and plan future financial actions. Machine learning and AI are well suited to manufacturing data. The manufacturing industry generates many analytical data that machines can analyze more easily. Machine learning models can predict the impact of individual variables in such complex situations, even for variables that are very difficult to interpret for humans. A lack of human capabilities prevents machines from being adopted in other industries involving language or emotions. COVID-19 also led manufacturers to become more interested in AI applications. The common use cases of AI in manufacturing are predictive maintenance, generative design, price forecasting of raw materials, robotics, edge analytics, quality assurance, inventory management, process optimization, product development, AI-power digital twin, design customization, performance improvement, and logistic optimization [66].

5.5.3 Healthcare and Life Sciences Applications

Human labor has traditionally been a big part of healthcare, but artificial intelligence is becoming an increasingly vital component. Artificial intelligence offers a wide range of healthcare services, including data mining, diagnostic imaging, medication management, drug discovery, robotic surgery, and medical imaging, to identify patterns and provide more accurate diagnoses and treatments. Technology giants like Microsoft, Google, Apple, and IBM significantly contribute to the healthcare sector.

It's unsurprising that artificial intelligence has a wide range of potential applications in the life sciences because they generate large amounts of data through experiments. This involves discovering and developing new drugs, conducting more efficient clinical trials, ensuring treatment is tailored to each patient, and pinpointing diseases more accurately.

In the information age, new technologies have affected many industries. Extensive use of artificial intelligence technology is reported by CB Insights in a 2016 report. A US $ 54 million investment in artificial intelligence is expected by 2020 by these companies. Here are a few examples of how artificial intelligence is impacting the healthcare industry today and in the future [67]:

- **Maintaining Healthcare Data:** Data management has become a widely used application of AI and digital automation in healthcare due to the necessity of compiling and analyzing information (including medical records). Using AI-based systems, data can be collected, stored, reformatted, and traced more efficiently and consistently.
- **Doing Repetitive Jobs:** AI systems can perform data entry, X-rays, CT scans, and other mundane tasks faster and more accurately. It takes a lot of time and resources to analyze data in cardiology and radiology. In the future, cardiologists and radiologists should only consider human monitoring in the most critical cases.
- **Design of Treatment Method:** The use of artificial intelligence systems helps physicians select the right, individually tailored treatment for each patient based on notes and reports in their patients' files, external research, and clinical expertise.
- **Digital Consultation:** AI-powered apps, such as Babylon in the United Kingdom, provide medical consultations based on a user's medical history and general knowledge of medicine. The app compares user symptoms with a database of illnesses using speech recognition. Babylon recommends actions based on the user's health history.
- **Virtual Nurses:** It is possible to monitor patients' health and follow treatments between doctor's visits with the help of startups that have developed digital nurses. Using machine learning, this program helps chronic illness patients. Parents of sick children can access basic health information and advice from Boston Children's Hospital's Alexa app. A doctor's visit is suggested based on symptoms; the app can answer questions about symptoms.
- **Medication Management:** An app created by the Patient Institute of Health monitors the use of patients' medications. Patients can automatically verify that they are taking their medications using the smartphone's webcam and artificial intelligence. Patients with serious medical conditions, patients who ignore doctors' advice, and clinical trial participants are most likely to use this app.
- **Drug Creation:** The development of CE actions requires billions of dollars and more than a decade of research. Increasing the speed and efficiency of this process can change the world. A computer program powered by artificial intelligence is being used to scan existing drugs in search of ones that can be redesigned to combat the Ebola virus. This type of analysis often takes months or years to find two actions that reduce the risk of Ebola infection in one day. This analysis typically takes months or years to discover a difference that can save thousands of lives.

- **Precision Medicine:** DNA information is used in genetics to find mutations and links to diseases. A body scan using AI can predict health risks based on people's genes and detect cancer and vascular diseases in advance.
- **Health Monitoring:** Heart rate and activity level can be monitored by wearable health trackers such as Fitbit, Apple, Garmin, and others. Physicians (and AI systems) can use this information to better understand patient needs and habits by sending alerts to users to exercise more.
- **Healthcare System Analysis:** Ninety-seven percent of healthcare invoices in the Netherlands are digital. Using artificial intelligence, a Dutch company identifies inefficiencies in treatment and workflow in healthcare systems to avoid unnecessary hospitalizations.

5.5.4 Telecommunication Applications

The telecommunications industry is highly complex and requires constant adjustments even though most of us take the Internet and communication access for granted. These needs can be met in several ways with the help of artificial intelligence. Using artificial intelligence in the telecom industry, rapid responses to customer inquiries can be automated, networks can be managed, and customized products can be designed. Telecom companies can benefit from AI by building stronger customer relationships and providing better services. Customers can obtain faster and smarter connections from many telecom operators in today's digital world. Telecommunications companies will acquire AI solutions that will guide many businesses toward success in the future. The main areas of this industry are as follows [68]:

- **Conversational Virtual Assistants:** Virtual assistants have been developed by many researchers for the telecom industry. Businesses and customers benefit from this technology by reducing expenses associated with customer service. Also, artificial intelligence has made it possible to interact with customers using natural speech processing. Artificial intelligence (AI) can solve various business problems by reducing human effort and creating a prosperous future.
- **Network Optimization:** The use of artificial intelligence has become increasingly popular among telecom businesses to improve their network infrastructure. AI enables network traffic management to benefit providers. In this way, network providers can predict and resolve issues before they occur. Also, AI monitoring systems can now track and trace the operations of telecom companies. Machine learning manages the data collected, and the network infrastructure is well-designed.
- **Predictive Maintenance:** Utilizing AI solutions to predict the future to manage malfunctions and resource utilization, predictive maintenance assists telecom

businesses in predicting the future cost-effectively. Artificial intelligence solutions can monitor complex communication hardware systems such as set-top boxes and cellphone towers to reduce operational costs and improve customer service. Artificial intelligence-powered drones have helped telecom operators capture and provide adequate resources during natural disasters by capturing damages to cell towers.

- **Robotic Process Automation (RPA):** Automating repetitive tasks with robots is very similar to automating business processes with artificial intelligence (AI) to improve operational efficiency. In addition to providing better customer service, RPA and AI innovations have improved workflow structures for managing sales orders, calls, emails, and psychographic profiling. This has led telecom companies to generate capital. Workers have also improved productivity and customer experience with AI-driven solutions for telecom businesses.

5.5.5 Oil, Gas, and Energy Applications

There is little room for error in the oil, gas, and energy sector because of safety and environmental concerns. Energy companies are turning to artificial intelligence to increase efficiency without incurring costs.

5.5.6 Aviation Applications

It is critical to use data effectively to optimize individual flights and the more comprehensive aviation infrastructure to maintain safe, efficient aviation, particularly in the context of rising fuel prices. This sector uses AI in the following ways:

- **Identify Routes with High Demand:** Providing enough flights between specific destinations while avoiding flying too many routes is crucial to maximizing profits while retaining customer loyalty. Airlines can use AI models to make informed decisions about route offerings based on factors like Internet traffic, macroeconomic trends, and seasonal tourism data.
- **Service to Customers:** The staffing capacity of most airlines is insufficient to handle individual customer queries and needs during major disruptions, such as those caused by massive weather events. AI is increasingly being incorporated into automated messaging to extract critical information from customer messages and respond accordingly. The customer may be directed to information about reporting lost luggage, for instance, if he or she inquires about their luggage.

5.6 Working Flow for AI-Powered Industry

The AI-powered industry has performed various steps to achieve business goals in any use case. These working steps for an AI-powered industry are shown in Fig. 5.4 and are described below:

- **Data Collection Process:** This is a very important and basic step for any industry to find appropriate data for specific use cases or projects. The data can be obtained from various sources such as publicly available platforms, collaborating with relevant authorities, etc. The data collection involves various steps, such as selecting, synthesizing, and sourcing the dataset.
- **Data Engineering and Model Development:** This second step is for product development and contains the process of data engineering and model development. In the data engineering process, various steps, such as data exploration, data clearing, data normalizing, feature engineering, and scaling, are performed to get suitable datasets for model development, while, in the model development process, model selection, model training, performance evaluation of the model, and model tuning are performed to get the correct trained model which can be used for developing a product.
- **Production:** In this step, the operation of the trained model is tested in various conditions to check its generalization usage and working ability in various conditions. This process contains various steps: registration, deployment, monitoring, and retraining.
- **Legal Constraints:** This is the most important process during product development using AI technology for any business case. This process contains various steps, such as legal and ethical approval, security, and product acceptance in terms of generalization.

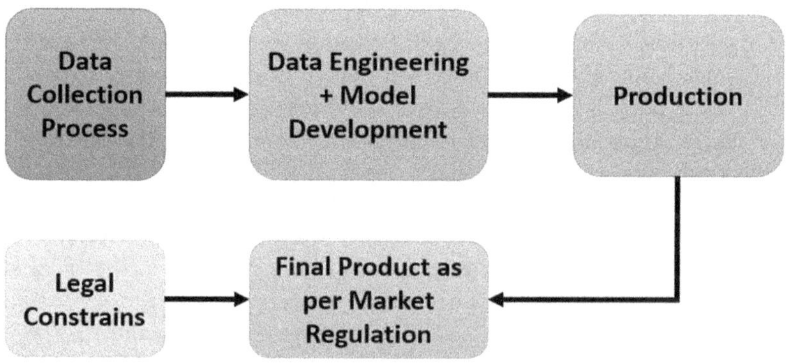

Fig. 5.4 Working flow of AI-powered industry

Once any AI-trained model or system fulfilled all required conditions for specific business use cases as a consumer product, the company can launch this model or system as a product that can be sold anywhere in the world.

References

1. C.M. Bishop, *Pattern Recognition and Machine Learning* (Springer International Publishing, Germany, 2006).
2. K.P. Murphy, *Machine Learning—A Probabilistic Perspective* (The MIT Press, Cambridge, 2012).
3. I. Goodfellow, Y. Bengio, A. Courville, *Deep Learning* (The MIT Press, Cambridge, 2016)
4. S.B. Kotsiantis, Supervised machine learning: a review of classification techniques. Informatica **31**, 249–268 (2007)
5. R. Thanki, S. Borra, Application of machine learning algorithms for classification and security of diagnostic images, in *Machine Learning in Bio-Signal Analysis and Diagnostic Imaging* (Academic Press, New York, 2019), pp. 273–292
6. A basic introduction to neural networks. http://pages.cs.wisc.edu/~bolo/shipyard/neural/local.html Accessed Feb 2018
7. S.S. Nath, G. Mishra, J. Kar, S. Chakraborty, N. Dey, A survey of image classification methods and techniques, in *2014 International Conference on Control, Instrumentation, Communication and Computational Technologies (ICCICCT)* (IEEE, 2014), pp. 554–557
8. S.D. Jawak, P. Devliyal, A.J. Luis, A comprehensive review of pixel-oriented and object-oriented methods for information extraction from remotely sensed satellite images with a special emphasis on cryospheric applications. Adv. Remote Sensing **4**(3), 177 (2015)
9. T. Lillesand, R.W. Kiefer, J. Chipman, *Remote Sensing and Image Interpretation* (Wiley, New York, 2014)
10. H. Zhang, The optimality of naive Bayes. AA **1**(2), 3 (2004)
11. V. Vapnik, *The Nature of Statistical Learning Theory* (Springer, New York, 1995)
12. C.W. Hsu, C.C. Chang, C.J. Lin, A practical guide to support vector classification (2016). https://www.csie.ntu.edu.tw/~cjlin/papers/guide/guide.pdf. Accessed Feb 2018
13. R. Saxena, How Decision Tree Algorithm Works (2017): https://dataaspirant.com/2017/01/30/how-decision-tree-algorithm-works/. Accessed Aug 2018
14. A.D. Kulkarni, A. Shrestha, Multispectral image analysis using decision trees. Int. J. Adv. Comput. Sci. Appl. **8**(6), 11–18 (2017)
15. A. Liaw, M. Wiener, Classification and regression by random forest. R news **2**(3), 18–22 (2002)
16. M.R. Segal, *Machine Learning Benchmarks and Random Forest Regression* (Kluwer Academic Publishers, Netherlands, 2004)
17. T.F. Cootes, M.C. Ionita, C. Lindner, P. Sauer, Robust and accurate shape model fitting using random forest regression voting, in *European Conference on Computer Vision* (Springer, Berlin, 2012), pp. 278–291
18. D.N. Kumar, Remote Sensing (2014). https://nptel.ac.in/courses/105108077/. Accessed July 2018
19. K. Wagstaff, C. Cardie, S. Rogers, S. Schrödl, Constrained k-means clustering with background knowledge, in *ICML*, vol. 1 (2001), pp. 577–584
20. J.A. Hartigan, M.A. Wong, Algorithm AS 136: a k-means clustering algorithm. J. R. Stat. Soc. C (Appl. Stat.) **28**(1), 100–108 (1979)
21. T. Kanungo, D.M. Mount, N.S. Netanyahu, C.D. Piatko, R. Silverman, A.Y. Wu, An efficient k-means clustering algorithm: analysis and implementation. IEEE Trans. Pattern Anal. Mach. Intell. **24**(7) 881–892 (2002)
22. K. Alsabti, S. Ranka, V. Singh, An efficient k-means clustering algorithm (1997)

23. A. Likas, N. Vlassis, J.J. Verbeek, The global k-means clustering algorithm. Pattern Recogn. **36**(2), 451–461 (2003)
24. L. Kaufman, P.J. Rousseeuw, *Finding Groups in Data: An Introduction to Cluster Analysis*, vol. 344 (Wiley, New York, 2009)
25. A.K. Jain, R.C. Dubes, Algorithms for clustering data (1988)
26. K. Mehrotra, C.K. Mohan, S. Ranka, *Elements of Artificial Neural Networks* (MIT Press, Cambridge, 1997)
27. I. Jolliffe, Principal component analysis, in *International Encyclopedia of Statistical Science* (Springer, Berlin, 2011), pp. 1094–1096
28. R.A. Schowengerdt, *Remote Sensing: Models and Methods for Image Processing* (Elsevier, Amsterdam, 2006)
29. R.C. Gonzalez, R.E. Woods, S.L. Eddins, *Digital Image Processing Using MATLAB*, vol. 624 (Pearson-Prentice-Hall, Upper Saddle River, 2004)
30. P. Comon, Independent component analysis, a new concept? Signal Process. **36**(3), 287–314 (1994)
31. X. Benlin, L. Fangfang, M. Xingliang, J. Huazhong, Study on independent component analysis application in classification and change detection of multispectral images. Int. Archiv. Photogramm. Remote Sensing Spatial Inform. Sci. **37**(B7), 871–876 (2008)
32. I. Dópido, A. Villa, A. Plaza, P. Gamba, A quantitative and comparative assessment of unmixing-based feature extraction techniques for hyperspectral image classification. IEEE J. Sel. Topics Appl. Earth Observ. Remote Sensor **5**(2), 421–435 (2012)
33. M.S.M. Al-Taei, A.H.T. Al-Ghrairi, Satellite image classification using moment and SVD method. Int. J. Comput. **23**(1), 10–34 (2016)
34. S. Brindha, Satellite image enhancement using DWT–SVD and segmentation using MRR–MRF model. J. Netw. Commun. Emerg. Technol. **1**(1), 6–10 (2015)
35. R.K. Jidigam, T.H. Austin, M. Stamp, Singular value decomposition and metamorphic detection. J. Comput. Virol. Hacking Techn. **11**, 203–216 (2015)
36. C. Biernacki, G. Celeux, G. Govaert, Assessing a mixture model for clustering with the integrated completed likelihood. IEEE Trans. Pattern Anal. Mach. Intell. **22**(7), 719–725 (2000)
37. C. Biernacki, G. Celeux, G. Govaert, Choosing starting values for the EM algorithm for getting the highest likelihood in multivariate Gaussian mixture models. Comput. Stat. Data Anal. **41**(3-4), 561–575 (2003)
38. Z. Zivkovic, Improved adaptive Gaussian mixture model for background subtraction, in *Proceedings of the 17th International Conference on Pattern Recognition, 2004. ICPR 2004*, vol. 2 (IEEE, 2004), pp. 28–31
39. C. Maugis, G. Celeux, M.L. Martin-Magniette, Variable selection for clustering with Gaussian mixture models. Biometrics **65**(3), 701–709 (2009)
40. G. McLachlan, D. Peel, *Finite Mixture Models*. Wiley Series in Probability and Statistics (2000)
41. T. Kohonen, Self-organized formation of topologically correct feature maps. Biol. Cybern. **43**(1), 59–69 (1982)
42. T. Kohonen, Analysis of a simple self-organizing process. Biol. Cybern. **44**(2), 135–140 (1982)
43. H. Ritter, T. Kohonen, Self-organizing semantic maps. Biol. Cybern. **61**(4), 241–254 (1989)
44. J.A. Kangas, T.K. Kohonen, J.T. Laaksonen, Variants of self-organizing maps. IEEE Trans. Neural Netw. **1**(1), 93–99 (1990)
45. E. Erwin, K. Obermayer, K. Schulten, Self-organizing maps: ordering, convergence properties and energy functions. Biol. Cybern. **67**(1), 47–55 (1992)
46. S. Kaski, T. Honkela, K. Lagus, T. Kohonen, WEBSOM—self-organizing maps of document collections. Neurocomputing **21**(1–3), 101–117 (1998)
47. M. Dittenbach, D. Merkl, A. Rauber, The growing hierarchical self-organizing map, in *Neural Networks, 2000. IJCNN 2000, Proceedings of the IEEE-INNS-ENNS International Joint Conference on*, vol. 6 (IEEE, 2000), pp. 15–19

48. D. Fumo, Types of Machine Learning Algorithms You Should Know (2017): https://towardsdatascience.com/types-of-machine-learning-algorithms-you-should-know-953a08248861. Accessed Mar 2020
49. Q-learning in python. https://www.geeksforgeeks.org/q-learning-in-python/. Accessed Mar 2020
50. R. Moni, (SmartLab AI), Reinforcement learning algorithms—an intuitive overview (2019): https://medium.com/@SmartLabAI/reinforcement-learning-algorithms-an-intuitive-overview-904e2dff5bbc. Accessed Feb 2023
51. A. Krizhevsky, I. Sutskever, G.E. Hinton, Imagenet classification with deep convolutional neural networks, in *Advances in Neural Information Processing Systems* (2012), pp. 1097–1105
52. S. Lawrence, C.L. Giles, A.C. Tsoi, A.D. Back, Face recognition: a convolutional neural-network approach. IEEE Trans. Neural Netw. **8**(1), 98–113 (1997)
53. H. Kandi, D. Mishra, S.R.S. Gorthi, Exploring the learning capabilities of convolutional neural networks for robust image watermarking. Comput. Secur. **65**, 247–268 (2017)
54. S.M. Mun, S.H. Nam, H.U. Jang, D. Kim, H.K. Lee, A robust blind watermarking using convolutional neural network (2017). ArXiv preprint arXiv: 1704.03248
55. A. Krizhevsky, I. Sutskever, G.E. Hinton, Imagenet classification with deep convolutional neural networks. Commun. ACM **60**(6), 84–90 (2017)
56. LeNet, http://deeplearning.net/tutorial/lenet.html. Accessed Feb 2019
57. Faster R-CNN, https://github.com/rbgirshick/py-faster-rcnn. Accessed Feb 2019
58. GoogLeNet, https://leonardoaraujosantos.gitbooks.io/artificial-inteligence/content/googlenet.html. Accessed Feb 2019
59. ResNet, https://github.com/gcr/torch-residual-networks. Accessed Feb 2019
60. R. Wang, T. Lei, R. Cui, B. Zhang, H. Meng, A.K. Nandi, Medical image segmentation using deep learning: a survey. IET Image Process. **16**(5), 1243–1267 (2022)
61. AI Research Trends (2016). https://ai100.stanford.edu/2016-report/section-i-what-artificial-intelligence/ai-research-trends. Accessed Feb 2023
62. P. Soni, Eight hot research domain topics in Artificial Intelligence (2020). https://er.yuvayana.org/8-hot-research-domain-topics-in-artificial-intelligence/. Accessed Feb 2023
63. Computer Vision Examples (2023). https://www.ibm.com/topics/computer-vision. Accessed Feb 2023
64. Personalized Recommendation Systems: Five Hot Research Topics You Must Know (2018). https://www.microsoft.com/en-us/research/lab/microsoft-research-asia/articles/personalized-recommendation-systems/. Accessed Feb 2023
65. Examples of Artificial Intelligence (AI) in 7 Industries (2022). https://emeritus.org/blog/examples-of-artificial-intelligence-ai/. Accessed Feb 2023
66. Cem Dilmegani (2022). Applications of AI in Manufacturing in 2023. https://research.aimultiple.com/manufacturing-ai/. Accessed Jan 2023
67. M. Kesavan, How Will Artificial Intelligence Reshape The Telecom Industry? (2022). https://itchronicles.com/artificial-intelligence/how-will-artificial-intelligence-reshape-the-telecom-industry/. Accessed Jan 2023
68. Swetha, 10 common applications of artificial intelligence in healthcare (2018). https://medium.com/artificial-intelligence-usm-systems/10-common-applications-of-artificial-intelligence-in-health-care-9d34ccccda5c. Accessed Apr 2020

Chapter 6
Advanced Technologies for Industrial Applications

Technology is rapidly evolving today, allowing for faster change and progress and accelerating the rate of change. Emerging technology is exciting, especially when it gives a field unexplored possibilities. Each year, new technologies grow more ubiquitous, and the digital tools that will be available in the future will be no exception. AI-as-a-Service and edge computing are only two examples of new tools that allow businesses to complete tasks in novel ways. The world is undergoing the fourth industrial revolution based on advanced technologies such as AI, machine learning, the Internet of Things, blockchain, etc. Researchers discovered ways to maintain and grow capacity, work securely, and meet the needs of important areas like medical devices and producing the ordinary things that keep the world running.

In many respects, technology-enabled manufacturers handle problems, with features like automation and remote monitoring and operation leading the way and allowing them to ramp up operations while keeping people safe and healthy.

In this chapter, we will discuss a few key and extremely valuable tools for industries in a number of different ways.

6.1 Industrial IoT (IIoT)

Industries are undergoing rapid technological transformations, particularly since the introduction of Industry 4.0, the fourth industrial revolution. Machines are now networked in a collaborative approach, with cyber-physical systems and analytical intelligence working together in a new way of production management, producing major industrial changes.

The current COVID-19 catastrophe, which has led to a global epidemic of life-threatening illnesses, has left over 800 million people without access to basic healthcare. Although there have reportedly been over 55,000 fatalities and an additional 2 million people are sent to hospitals each week, this issue is not going

R. Thanki, P. Joshi, *Advanced Technologies for Industrial Applications*,
https://doi.org/10.1007/978-3-031-33238-8_6

away soon. Mobile applications, robots, Wi-Fi cameras, scanners, and drones are being used to stop the spread of viruses. The significant contribution of digital technology to pandemic control has been impacted by Industry 4.0.

However, due to the diverse range of Industry 4.0 technologies, such as mobile, cloud computing, Big Data, analytics tools, machine-to-machine (M2M), 3D printing, and robots, the route to digital transformation is not easy, although these were some of the technologies that sparked Industry 4.0 much broader.

The term "Industrial Internet of Things" was created to explain the Internet of Things (IoT) as it is used in a variety of industries, including manufacturing (Industry 4.0), logistics, oil and gas, transportation, energy/utilities, mining and metals, aviation, and other industrial sectors, as well as the use cases that are specific to these industries.

These technologies are part of the industrial Internet of Things (IIoT), one of the most well-known technology concepts. In an industrial context, the IIoT is a physical network of things, objects, or devices (that have embedded technology) for sensing and remote control, allowing better integration between the physical and cyber worlds.

6.1.1 Internet of Health Things

To create an even more complicated system that includes people, robots, and machines, the Internet of Everything (IoE) generalizes machine-to-machine (M2M) connections for the Internet of Things (IoT). IoT in healthcare is generally called IoHT (Internet of Healthcare Things) or the Internet of Medical Things (IoMT). IoHT primarily focuses on the wireless connecting of a network of medical equipment and body sensors with the cloud to gather, analyze, organize, and process health data. The proper protocols are used for secure connections and effective machine-to-machine data transfer by health devices that collect data in real time.

The Internet of Things (IoT) is a network of smart devices, wireless sensors, and systems that combine several recent technological advancements. In the healthcare industry, low-power, low-latency technologies are in high demand. With the development of wireless communications, the network structure has significantly changed in this time period. Additionally, some research investigates how the Internet of Things (IoT) will support the next-generation network architecture, indicating how embedded devices can quickly connect with one another. The configuration of wireless and low-power, low-latency medical equipment for IoT devices will fundamentally alter healthcare.

Continuous monitoring of the health of an unexpectedly large number of patients throughout both the pre- and post-infection stages is highly essential during the COVID-19 pandemic. Both carers or healthcare practitioners and patients have effectively adopted remote patient monitoring, screening, and treatment via telehealth facilitated by the Internet of Health Things (IoHT). Smart devices powered by the Internet of Things (IoHT) are proliferating everywhere, especially in the midst

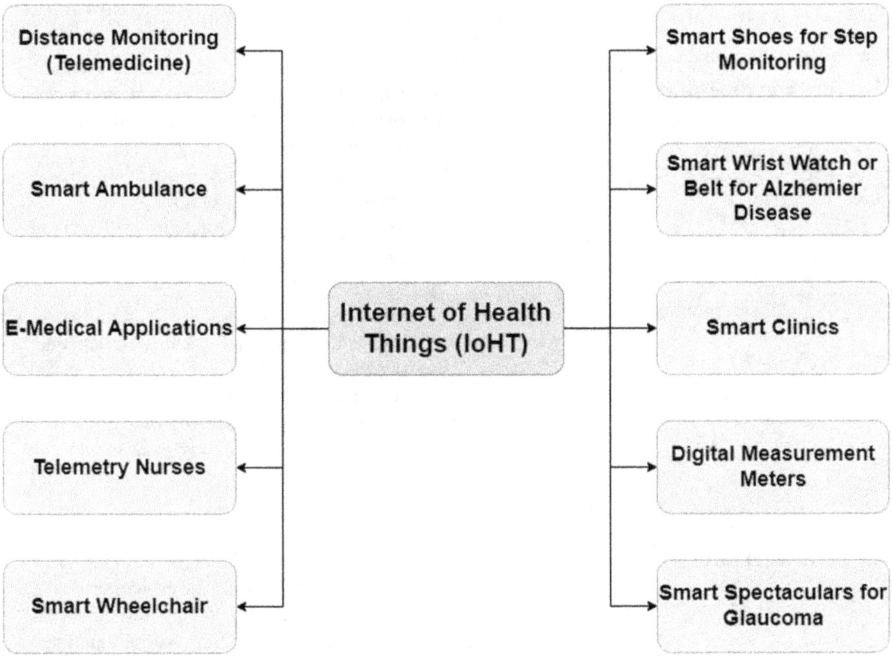

Fig. 6.1 Examples of Internet of Health Things (IoHT) applications

of a global epidemic. However, healthcare is seen as one of the IoT's most difficult application sectors due to a large number of needs. The graphical overview of IoHT is shown in Fig. 6.1.

6.1.1.1 Recent Case Study and Enabling Technologies Overview

IoHT has a great ability to produce good results with the aid of cutting-edge technologies. In the field of medicine, it has become a new reality of an original concept that offers COVID-19 patients the greatest care and conducts accurate surgery. During the ongoing pandemic, complicated situations are readily managed and controlled digitally. IoHT takes on fresh issues in the medical industry to develop top-notch assistance programs for physicians, surgeons, and patients. To implement IoHT successfully, certain process steps are carefully identified. These steps included the setup of networking protocols and the acquisition of sensor data with a secured transmission system.

IoHT integrates machines, tools, and medical supplies to produce intelligent information systems that are tailored to the needs of each COVID-19 patient. An alternative interdisciplinary strategy is required to maximize output, quality, and understanding of emerging diseases. IoHT technology tracks change in critical patient data to get pertinent data. The various IoHT technologies that were useful in healthcare during the COVID-19 pandemic are covered in Table 6.1.

Table 6.1 Emerging technologies for implementation of Internet of Health Things

Sr. no.	Technologies	Description
1	Fog computing	– Fog computing is a type of computing architecture where a network of nodes continuously receives data from IoHT devices – With millisecond response times, these nodes process data in real time as it comes in – The nodes periodically transmit to the cloud analytical summary data
2	Edge computing	– Edge computing enables IoHT data to be acquired and processed locally rather than sending the data back to a data center or cloud – Edge computing is a strategy for computing on location where data is received or used – A potent technique to quickly examine data in real time is by combining edge computing and IoT
3	Wireless/node based	The upcoming technology, the cloud computing solution known as Function-as-a-Service, or FaaS, enables developers to create, launch, and manage application packages as functions without having to maintain their infrastructure
4	ML/AI	Machine learning (ML), deep learning (DL), traditional neural networks, fuzzy logic, and speech recognition are only a few of the subsets of artificial intelligence that have distinct skills and functions that can enhance the performance of contemporary medical sciences
5	Digital twin	A digital twin is a representation of a physical product, procedure, or service in the digital world
6	Bluetooth Low Energy (BLE)	The Bluetooth Special Interest Group created and promoted Bluetooth Low Energy, a wireless personal area network technology with new uses in the home entertainment, fitness, security, and healthcare sectors

6.2 Autonomous Robots

Robotics is the study of robots, which are electromechanical devices used for various tasks. Due to their popularity and ability to accomplish activities that people cannot do, the most common robots are used in dangerous environments. For the past 15–20 years, the most common uses of robotics have involved basic industrial or warehouse applications or teleoperated mobile robots with cameras to see objects out of reach. For instance, flying robots (also known as drones) are used for disaster response, in addition to automated guided vehicles (AGVs) for material movement in factories and warehouses, and underwater robots are employed to search for and find shipwrecks in the deepest parts of our oceans.

Despite the fact that using robots in this way has been very successful over the years, the usage of completely autonomous robots is in no way represented by these examples. Some robots can do activities independently, while others require human assistance to complete or direct them. Robots can be employed in various industries, including the medical area, military applications, and space communication.

Based on the control system it features, an automatic robot is a sort of manipulated robotic system regarded as one of the early robotic systems. Automatic robots are categorized into *four basic groups* based on their traits and intended uses.

1. Programmable robots
2. Non-programmable robots
3. Adaptive robots
4. Intelligent robots (*collaborative robots and soft robots*)

Here, in the next sections are surveys of use cases of intelligent robots: cobots and soft robots. The main goal of this chapter is to focus on the advanced technologies used in different industrial domains. Many businesses and organizations comprise the automotive industry, which aims to design, develop, market, manufacture, and sell automobiles using human interface robots.

6.2.1 Collaborative Robots (Cobots)

A type of robotic automation known as collaborative robots is designed to function securely alongside human workers in a shared, collaborative workspace. In most situations, a collaborative robot handles monotonous, menial chores, while a human employee handles more difficult, thought-intensive jobs. Collaborative robots are accurate, reliable, and repeatable to supplement human workers' intelligence and problem-solving abilities.

The notion of Industry 4.0 includes collaborative robots and how they might be used in assisted assembly. They can do simple manipulations or boring, repetitive assembly activities in the same workstation as human workers. Due to the small number of actual applications in production processes at the moment, this area is still open to fresh research, methodological development, and determination of fundamental needs.

The *primary benefit* of deploying collaborative robots during assembly is a short transfer time for parts between manual and automated activity. Some advantages include a built-in vision system for further manual operation inspection, the ability to give an interface for digital data collecting from sensors, and connectivity to external cloud platforms.

Compared to industrial robots, collaborative robot designs are very different. Collaborative robots are, first and foremost, made for safety, with rounded edges, force constraints, and light weights. Most collaborative robots are outfitted with various sensors to prevent accidents with human employees and safety mechanisms to shut down in the event of any unintended contact.

Industrial and collaborative robots may completely automate a process without involving people, whereas cobots collaborate with humans. This is the main distinction between the two types of robots. Additionally, cobots cannot accomplish some demanding manufacturing tasks that an industrial robot needs to handle [1].

6.2.2 Soft Robotics

In the subject of robotics, known as "soft robotics," compliant materials, as opposed to rigid linkages, are used to design, manufacture, and control robots. The soft robotics name itself suggests not in terms of *soft* but movable and adjustable. In Industry 5.0, it is clear that people have smart everything with the Internet of Everything (IoE), and the soft robotics will also play a vital role.

In Healthcare 4.0, many recent cases have been authored with wearable and haptic vibrotactile devices [2]. However, this enhancement still lacks communication with end-to-end latency and reliability. For simplicity, it has been assumed that augmented reality and virtual reality-based robotics have stability and transparency, but the M2M interface and E2E latency still need to be improvised. Those issues and gaps can be sorted out via 6 G-enabled technologies because, somehow, 3GPP architecture is adopted via 6 G-enabled networking technologies.

In automotive industries, various arm-based robots have been implemented with different applications, for instance, car manufacturing, drone manufacturing, agriculture-based manufacturing, and many more. It is also possible that cobots are used in some applications. Still, soft robotics with multiple degrees of freedom (DoF) has been included in operations where human operators do not require it. For instance, **FANUC** robots have multiple use cases which are easy to use and sufficient for modest powder-coating operations while being delicate enough for vehicle painting. Finally, their mid-range arms may do anything from pick-and-place to welding and machine tending.

Physical treatment and rehabilitation have been investigated using soft robots in healthcare and medical applications [2]. Apart from that, teleoperations and telerobotic surgeries have been implemented using robotic simulators, which are also robust and feasible. Many comparative studies show that blood loss during teleoperations is lower than performed by human operators. Figure 6.2 examines how seniors with moderate cognitive impairment [3] interact with humanoid robot-accessed serious games as part of a cognitive training program.

Different robots and their specific use have been shown in Table 6.2.

Soft robots find it challenging to provide sensory input [1] since deformation can occur in any direction. The difficult part is deciding which parameters to measure and how to measure them. For soft robots, visual feedback is one possible sensing

Fig. 6.2 Example of soft robotics (Source: Humanoid robot on mild cognitive impairment older adults)

Table 6.2 Different soft robots with their specific use case

Sr. no.	Type of soft robot	Purpose of soft robot
1	Soft robotic catheters	Navigate complicated, curved blood tubes or other bodily par
2	Soft robotic exoskeletons	Help mobility-impaired people recover
3	Soft robotic prosthetic devices	To be more flexible and patient-friendly than typical prosthetic systems
4	Soft robotic endoscopes	Explore complicated bodily cavities
5	Octobot	Underwater exploration and monitoring
6	Soft robotic puppets	For entertainment purposes, such as in theme parks or interactive exhibits

Table 6.3 KPIs for soft robots and target audience

KPI	Robot	Task of robot	Patients	Year of study
Position accuracy	ALTER-EGO	Augmented reality/visual reality	Elders	2019
Latency	NAO, Qbo, and Hanson robot	Live streaming	Elders and kids	2017
Retainability (Total operation time)	Doro	Reminder action and medical guidance	Elders with chronic issues [20]	2017
Throughput	Domestic Doro	Augmented reality	Elders	2016
Reliability or workload	Cloud robot [21, 22]	Monitoring important indication	Elders	2015

method. The resulting motion may be observed on an external picture or video capture device. A tactile deformable sensor offers a potential means of obtaining sensory feedback.

NASA set an example of teleoperation in space to solve the technical problem in the robotic models [19]. Telerobots are also used in maritime applications to study and observe marine life. However, soft robots have a few challenges and issues as shown in Table 6.3.

Current rehabilitation [4–6], surgical, and diagnostic soft robot concepts are grouped by application and appraised for functionality.

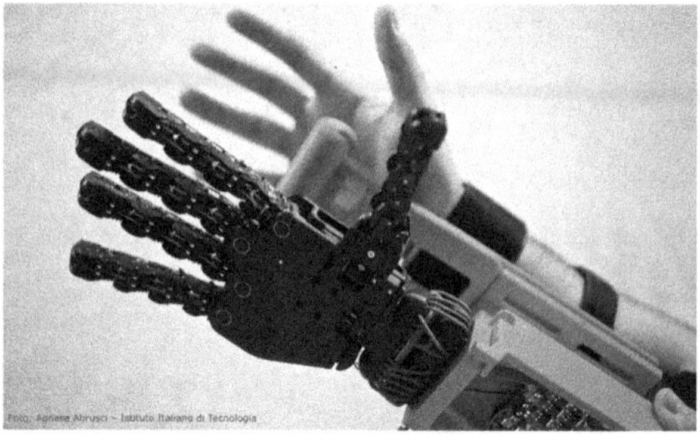

Fig. 6.3 Soft robotic hand with smooth fingertips (Source: Recent research studies in Istituto Italiano di Tecnologia, Genoa, Italy)

Fig. 6.4 A fully automotive robot with soft arm (Source: Recent research studies in Istituto Italiano di Tecnologia, Genoa, Italy)

The first prototype of the Pisa/IIT SoftHand [7], a highly integrated robot hand with a humanoid shape, robustness, and compliance, is displayed and described. Extensive grab cases and grasp force measurements finally support Fig. 6.3's hand.

A dual-arm mobile platform called ALTER-EGO [8], designed by its authors using soft robotic technology for the actuation and manipulation layers, is shown in Fig. 6.4. The flexibility, adaptivity, and robustness of this type of technology's features enable ALTER-EGO to interact with its surroundings and objects and improve safety when the robot is near people.

6.3 Smart and Automotive Industries

In today's world, automation and control have replaced human work, which is also an advancement toward smarter cities. The automotive industry is one of the reputed examples of smart manufacturing units. Various organizations and companies provide smart manufacturing in automotive industries by giving robotic solutions and decreasing manpower in field sites.

ABB is a Swiss company with over 130 years of technical innovation. ABB is a pioneer in Industry 4.0 and a leader in industrial digitization today. Robots made by ABB are robust, adaptive, and versatile thanks to their single- and dual-arm designs. An extensive selection of industrial robots is available from **KUKA**. No matter the application's difficulty, you will always discover the appropriate one. Also, **Rockwell** automation provides feasible and efficient solutions for various automotive industries.

Industrial robotics is becoming more and more prevalent due to their effectiveness and precision, especially in the manufacturing business, even though full automation and the employment of robots in a residential setting are still the exceptions rather than the rule. The Statista Technology Market Forecast predicts that by 2021, over 500,000 industrial robot systems will be in use worldwide. According to Fig. 6.5, based on the corresponding dataset, sales of robots targeted toward two industries in particular account for the largest portion of total revenue.

By introducing edge computing and beyond 5G networking, it becomes easy to be everything available at remote sites faster and with the minimum end-to-end transmission delay. The Internet of Things enables better transportation efficiency, cutting-edge vehicle management capabilities, and a superior driving experience in the automotive sector, paving the path for autonomous cars, which were formerly thought to be a future vision.

More complicated improvements will become available as embedded vehicle IoT systems develop. Additionally, the ability of linked car technology and the speed at which mobile communications develop allow automakers to keep introducing fresh and intriguing services.

6.4 Human and Machine Interfacing (HMI)

HMIs are user interfaces or dashboards that connect people to machines, systems, and devices. HMI is most commonly used in industrial processes for screens that allow users to interact with devices. Graphical user interfaces (GUIs) and human-machine interfaces (HMIs) have some similarities but differ. GUIs are often used

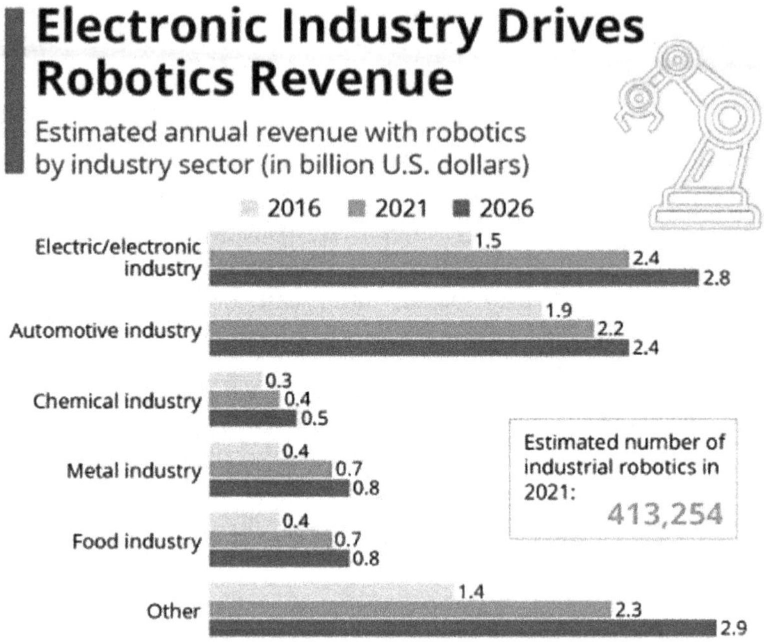

Fig. 6.5 Industrial robot revenues. Source: Statista Technology Market Outlook

for visualization in HMIs. The main purpose of HMI examples is to provide insight into mechanical performance and progress, regardless of the format or the term you use to refer to them. HMIs can be used in industrial settings to:

- **Data visualization.**
- **You can track the time, trends, and tags associated with your production.**
- **Monitor key performance indicators.**
- **Monitor the outputs and inputs of the machine.**

Most industrial organizations use HMI technology to interact with their machines and optimize their industrial processes. Operators, system integrators, and engineers use HMIs the most, especially control system engineers [9]. For these professionals, HMIs are essential tools for reviewing and monitoring processes, diagnosing problems, and displaying data. HMI is used in the following industries:

- **Energy and power**
- **Food**
- **Manufacturing and production**
- **Gas and oil**
- **Transportation**
- **Water processing**
- **And many more**

Technology developments in HMI have occurred due to changing business and operational needs in the past decade. Human-machine interfaces are becoming increasingly evolved as time passes. Many types of HMIs are available as in current AI world, including traditional, high-performance and touch screens. Modernizing equipment interfaces allows us to interact with it and analyze it more effectively.

6.5 AI Software

Artificial intelligence software enables humans to employ artificial intelligence (AI) to process large quantities of data to solve tasks that would otherwise require human intelligence. Such tasks include image, video, voice, text, and natural language processing. There is exponential growth in the strategic importance of artificial intelligence across a wide range of industries. Many businesses are exploring and investing in AI solutions to stay competitive [10, 11].

- **Viso Suite Platform:** It is the only comprehensive platform for computer vision applications available worldwide. It enables AI vision applications to be developed, deployed, scaled, and secured using software infrastructure. Computer vision applications from the world's largest companies are delivered and maintained using the Viso platform. By integrating Viso Suite, you can avoid spending time and money on integrating point solutions for each computer vision project. The platform supports a complete AI lifecycle, including data collection, image annotation, model training, application development, deployment, configuration, and monitoring.
- **Content DNA Platform:** The Content DNA software platform specializes in video content analysis using artificial intelligence. The product performs video-related tasks for broadcasters and telecom companies, including scene recognition, anomaly detection, and metadata enrichment. No matter your background, you can learn and use the platform easily.
- **Jupyter Notebooks:** Code-first users can write and run computer code using Jupyter Notebooks, an open-source software. As the name suggests, Jupyter supports Julia, Python, and R as its three core programming languages. The Notebook allows you to run code cells and see the output without writing extra code. Due to these advantages, Jupyter Notebooks are popular for developing AI applications, exploring data, prototyping algorithms, and implementing vision pipelines.
- **Google Cloud AI Platform:** It offers many machine learning tools. Google Cloud Platform (GCP) is one of the most popular platforms for scientists and developers. Machine learning projects can be performed more quickly and efficiently using the AI software tools Google Cloud provides. ML applications related to computer vision, translation, natural language, and video can be built using pre-trained cloud APIs. PyTorch, TensorFlow, and scikit-learn are the open-source frameworks that Google Cloud supports.

- **Azure Machine Learning Studio:** You can create and deploy robust machine learning models with the Azure Machine Learning Studio. TensorFlow, PyTorch, Python, R, and other open-source frameworks and languages are among those supported by the platform. A wide range of users, including developers and scientists, can benefit from Microsoft AI software.
- **Infosys Nia:** Businesses and enterprises can simplify AI implementation with Infosys Nia, an AI software platform. A wide range of tasks is possible with it, such as deep learning, natural language processing (NLP), data management, etc. Companies can automate repetitive tasks and schedule responsibilities with AI on existing Big Data using Infosys Nia. Thus, organizations can be more productive, and workers can accomplish their tasks more efficiently.
- **Salesforce Einstein:** Businesses can use Salesforce Einstein to build AI-enabled applications for their customers and employees with Salesforce's analytics AI platform for CRM (customer relationship management). Predictive models can be built using machine learning, natural language processing, and computer vision. Model management and data preparation are not required with artificial intelligence tools.
- **Chorus.ai:** Specifically designed for sales teams on the verge of growth, Chorus.ai offers conversation intelligence features. The application assists you in recording, managing, and transcribing calls in real time and marking important action items and topics.
- **Observe.AI:** With Observe.AI, businesses can transcribe calls and improve performance by using automated speech recognition. User-friendly automation tools are available in both English and Spanish. Using the most recent speech and natural language processing technology allows businesses and organizations to analyze calls effectively. Other business intelligence tools can also be integrated with the tool.
- **TensorFlow 2.0:** For developers, TensorFlow (TF) is an open-source machine learning and numerical computation platform based on Python. Artificial intelligence software TensorFlow was created by Google.
- **H2O.ai:** Businesses can easily train ML models and apps with H2O.ai, an end-to-end platform. Using AutoML functionality, beginners and experts can create or train AI models. Besides tabular data, the platform can handle text, images, audio, and video files. Businesses can manage digital advertising, claims management, fraud detection, and advanced analytics and build a virtual assistant with the open-source machine learning solution for enterprises.
- **C3 AI:** As a provider of AI SaaS (software as a service), C3 AI provides AI software as a service (SaaS) to accelerate digital transformation and build AI applications. The C3 AI Suite and C3 AI applications are available from C3.ai as software solutions for artificial intelligence. This AI platform company offers a variety of commercial applications, including energy management, predictive maintenance, fraud detection, anti-money laundering, inventory optimization, and predictive CRM.
- **IBM Watson:** Using IBM Watson, companies and organizations can automate complex machine learning processes, predict future results, and optimize

employee productivity. To make sense of data, recognize patterns, and predict the future, IBM offers a broad AI portfolio that includes pre-trained models and custom machine learning models.

- **DataRobot:** Organizations can fast-track the development of predictive models and uncover insights using DataRobot's automated machine learning platform. This tool can create and deploy machine learning models quickly and efficiently for data scientists, developers, and business analysts.
- **Tractable:** Tractable is an AI-driven platform that offers automated and efficient accident assessment solutions to the automotive, industrial, and insurance industries. As a result, it is easier to assess damaged vehicles, claims are processed faster, and operations are streamlined.
- **Symantec Endpoint Protection:** A company's cybersecurity needs to be evaluated if it conducts any part of its business online. Symantec Endpoint Protection uses machine learning technology to secure digital assets. In time, the program can learn to distinguish between safe and malicious files on its own as it encounters different security threats. Symantec's website explains that the platform's AI interface can alleviate the need to configure software and run updates manually by automating updates and learning from security threats.
- **Outmatch:** Using AI-enabled technology, Outmatch aims to streamline the entire recruiting process. Recruiting teams can reduce spending by up to 40% using Outmatch's AI-enabled hiring workflow. Using Outmatch's tools, users can schedule interviews, check references, and screen candidates behaviorally and cognitively.
- **Tableau:** Business strategy and industry forecasts can be developed using Tableau's visualization software. Users access data insights faster with Tableau's AI and augmented analytics features than they would through manual methods.
- **Oracle AI:** Developed specifically for developers and engineers, Oracle AI analyzes customer feedback and creates accurate predictive models using the extracted information. According to the company's website, developers do not have to create applications from scratch with Oracle's platform because it automatically pulls data from open-source frameworks. A chatbot tool on its platform connects customers with appropriate resources or support based on their needs.
- **Caffe:** It is an open-source framework for defining, designing, and deploying machine learning applications. Caffe is a digital project launcher developed by Berkeley AI Research, incorporating Python for modeling, testing, and automatically resolving bugs.
- **SAS:** SAS is a data management system based on open-source and cloud-enabled technologies that help businesses grow and progress. According to SAS's website, the platform can help companies better control their direction through customer intelligence, risk assessment, identity verification, and business forecasting.
- **Theano:** Developers can use Theano to successfully create, optimize, and launch code projects using an AI-powered library that integrates with Python. According

to the product's website, Theano uses machine learning to diagnose bugs and fix malfunctions independently, with minimal support from outside.

- **OpenNN:** OpenNN, an open-source software library that uses neural network technology, can interpret data more quickly and accurately. OpenNN claims to be faster than its competitors at analyzing and loading massive datasets and training models, according to its website.
- **Tellius:** Tellius, an AI-driven software, helps businesses better understand their strategies, successes, and growth opportunities. Using Tellius's platform, employees can access an intelligent search function that organizes data and makes it easier to understand. Their business outcomes can be analyzed and understood through this process.
- **Gong.io:** Gong.io, an AI-driven platform, analyzes customer interactions, forecasts future deals, and visualizes sales pipelines.
- **Zia by Zoho:** With Zoho's Zia, companies can gather organizational knowledge and turn customer feedback into strategies using a cloud-based AI platform. According to Zia's website, its AI tools can analyze client schedules, sales patterns, and workflow patterns to improve employee performance.
- **TimeHero:** Users can manage their projects, to-do lists, and schedules using TimeHero's AI-enabled time management platform. According to TimeHero's site, the platform's machine learning capabilities can notify employees when meetings occur, when emails are due, and when projects are due.

6.6 Augmented and Virtual Reality (AR/VR)

Augmented reality (AR) and virtual reality (VR) are two different technologies used to enhance the experience of interacting with the digital world. Augmented reality (AR) is a technology that overlays digital information in the real-world environment. This technology can be experienced through smartphones, tablets, smart glasses, or headsets. It enhances real-world experiences by adding digital elements such as images, videos, sounds, and 3D models to the physical world.

Virtual reality (VR) is a technology that creates a simulated, computer-generated environment that can be experienced through a headset or a display. VR immerses the user in an artificial environment, creating the illusion of being in a different world. The user can interact with the environment and other objects through physical movements and controllers. Augmented and virtual reality (AR/VR) have significant potential to transform numerous industries and bring about innovative solutions to challenges faced by various sectors. Here are some examples of how AR/VR is being used in different sectors:

1. **Education:** AR/VR can create immersive and interactive learning environments that enhance student engagement and knowledge retention. For example, VR can be used to simulate scientific experiments or historical events, while AR can be used to provide real-time feedback to students during a lesson.

2. **Healthcare:** AR/VR is used in medical training to simulate surgical procedures, anatomy, and patient diagnosis. VR can also be used to manage pain and help patients overcome phobias or anxiety.
3. **Retail:** AR/VR is being used to enhance customer experiences, such as allowing customers to virtually try on clothing or see how furniture would look in their homes before making a purchase.
4. **Manufacturing:** AR/VR can assist with product design and prototyping, allowing engineers to visualize and test designs in a virtual environment before manufacturing.
5. **Construction:** AR/VR can be used in architecture and construction to visualize and plan projects, allowing designers to create virtual models of buildings and test different materials and designs.
6. **Entertainment:** AR/VR creates immersive gaming experiences, allowing players to interact with virtual environments and objects.
7. **Military and Defense:** AR/VR is used for training simulations and providing real-time information to field soldiers.

Overall, AR/VR has the potential to revolutionize the way various industries operate, improving efficiency, reducing costs, and enhancing the overall customer experience. Various methods are used in augmented and virtual reality to create immersive experiences. Here are some of the most common methods:

1. **Marker-Based AR:** This method uses markers or triggers, such as QR codes or images, to create augmented reality experiences. When the device's camera is pointed at the marker, the app overlays digital information on top of the real-world image.
2. **Location-Based AR:** This method uses the device's GPS to create location-based augmented reality experiences. For example, an AR app can provide information about a historical landmark or provide directions to a nearby store.
3. **Projection-Based AR:** This method projects digital information onto real-world surfaces, such as walls or floors, to create an augmented reality experience.
4. **Head-Mounted Displays (HMDs):** HMDs, such as VR headsets, are worn on the head and immerse the user in a virtual reality experience by displaying computer-generated graphics in front of their eyes.
5. **Room-Scale VR:** This method uses multiple sensors and cameras to create a virtual reality experience that allows the user to move and interact with objects in a designated physical space.
6. **360-Degree Video:** This method captures video from all angles and allows the user to view the video in a 360-degree immersive environment.
7. **Hand Tracking and Controllers:** This method uses hand tracking technology or handheld controllers to allow users to interact with digital objects in a virtual environment.

These are just some of the methods used in augmented and virtual reality. Technology constantly evolves, and new methods are being developed to create even more immersive experiences.

6.7 Blockchain and Cybersecurity

Blockchain is a distributed ledger technology that allows for secure and transparent transactions between parties without needing a trusted intermediary. Conversely, cybersecurity protects computer systems and networks from unauthorized access, theft, damage, and other threats. Blockchain technology has a significant potential in the realm of cybersecurity. Here are some ways that blockchain can enhance cybersecurity:

1. **Immutable Record-Keeping:** Blockchain technology's decentralized and immutable nature makes it difficult to tamper with data stored on a blockchain, providing greater security and integrity.
2. **Cryptographic Security:** Blockchain uses cryptography to secure transactions and protect sensitive data. The use of cryptographic techniques can make it difficult for attackers to access or manipulate data.
3. **Distributed Security:** Because blockchain is a distributed technology, there is no central point of failure, and the network is less vulnerable to hacking or other forms of cyber-attacks.
4. **Decentralized Identity Management:** Blockchain technology can be used for decentralized identity management, where individuals can control their personal data and authenticate their identity without relying on a central authority.
5. **Smart Contract Security:** Smart contracts, which are self-executing contracts with the terms of the agreement written into code, can be used to automate transactions and reduce the risk of fraud or human error.

Overall, blockchain technology can provide greater security and transparency in the realm of cybersecurity. By creating a decentralized and secure environment, blockchain has the potential to reduce the risk of cyber-attacks and protect sensitive data. However, it is important to note that blockchain technology is not a panacea for all cybersecurity issues, and proper implementation and management are essential to ensure its effectiveness.

Various methods are used in blockchain and cybersecurity to ensure the security and integrity of data stored on a blockchain network. Here are some of the most common methods:

1. **Encryption:** Encryption is the process of converting plain text data into a coded form that can only be read by authorized parties. This technique is commonly used in blockchain to protect sensitive data.
2. **Hashing:** Hashing transforms data into a fixed-length string of characters representing the original data. This technique is used in blockchain to create a unique identifier for each data block, ensuring the data's integrity.
3. **Digital Signatures:** Digital signatures are used to authenticate the sender's identity of a message or transaction. In the blockchain, digital signatures are used to ensure that only authorized parties can access and modify data stored on the blockchain.

4. **Multi-factor Authentication:** Multi-factor authentication requires users to provide two or more forms of identification to access a system or network. This technique can be used to protect access to blockchain networks and ensure that only authorized users can access the data.
5. **Consensus Mechanisms:** Consensus mechanisms are used in blockchain to ensure that all parties on the network agree on the validity of transactions and data stored on the blockchain. There are several consensus mechanisms, including proof of work and proof of stake.
6. **Firewalls and Intrusion Detection Systems:** Firewalls and intrusion detection systems are used to protect computer systems and networks from unauthorized access and cyber-attacks. These techniques can be used in conjunction with blockchain technology to provide greater security.
7. **Penetration Testing:** Penetration testing involves testing the security of computer systems and networks by attempting to exploit vulnerabilities. This technique can be used to identify and address security weaknesses in blockchain networks.

These are just some of the methods used in blockchain and cybersecurity. The use of these techniques and others can help ensure the security and integrity of data stored on blockchain networks.

Blockchain technology and cybersecurity are significant because they provide enhanced security and integrity for digital transactions and data. Here are some of the main reasons why blockchain and cybersecurity are important:

1. **Security:** Blockchain technology provides a secure and transparent environment for digital transactions and data storage. Because of its decentralized nature, it is more difficult for hackers to compromise the network, steal data, or disrupt transactions.
2. **Transparency:** Blockchain provides transparency by allowing all parties on the network to view and verify transactions. This transparency can help prevent fraud and increase trust in transactions.
3. **Immutability:** Blockchain is immutable, meaning that once data is stored on the network, it cannot be changed or deleted. This feature provides greater security and helps prevent data tampering.
4. **Decentralization:** Blockchain is a decentralized technology, meaning no central authority controls the network. This feature provides greater security by eliminating single points of failure and reducing the risk of cyber-attacks.
5. **Trust:** Blockchain technology can help build trust between parties in digital transactions by providing a secure and transparent environment for exchanging data and assets.
6. **Cost Savings:** Blockchain technology can reduce costs by eliminating the need for transaction intermediaries and reducing the risk of fraud and other forms of financial loss.
7. **Innovation:** Blockchain technology can drive innovation by enabling new business models and providing a platform for developing new applications and services.

Overall, blockchain and cybersecurity are significant because they provide enhanced security, transparency, and trust in digital transactions and data storage. By leveraging these technologies, businesses and organizations can reduce costs, increase efficiency, and drive innovation. Blockchain technology and cybersecurity have many potential application areas across various industries. Here are some examples:

1. **Financial Services:** Blockchain technology can be used in the financial industry to securely store and transfer digital assets, such as cryptocurrencies, and streamline payment processing and settlement.
2. **Supply Chain Management:** Blockchain technology can be used in supply chain management to track and verify the authenticity of products as they move through the supply chain, providing greater transparency and reducing the risk of counterfeit products.
3. **Healthcare:** Blockchain technology can be used in healthcare to securely store and share patient data, ensuring patient privacy and facilitating the exchange of medical records between healthcare providers.
4. **Voting:** Blockchain technology can be used in voting systems to increase transparency and security in elections by creating a tamper-proof record of votes and preventing voter fraud.
5. **Intellectual Property:** Blockchain technology can be used to protect intellectual property rights by providing a secure and transparent platform for registering and tracking patents, copyrights, and other forms of intellectual property.
6. **Cybersecurity:** Blockchain technology can enhance cybersecurity by creating a secure and decentralized platform for storing and sharing sensitive data and providing a tamper-proof record of cyber-attacks and security breaches.
7. **Energy Management:** Blockchain technology can be used in energy management to facilitate peer-to-peer energy trading and securely manage energy supply and demand.

These are just a few examples of blockchain technology and cybersecurity application areas. As these technologies continue to evolve and mature, we can expect to see even more innovative use cases across various industries.

6.8 Challenges and Open Research Problems in Various Domains

Researchers can work together as a community to address some of the biggest challenges and opportunities in real-world problems in the coming years [12]. Here are a few of the open problems that the participants highlighted: machine learning, medical imaging, natural language processing, robotics, and wireless communications.

6.8.1 Machine Learning

Deep learning was first applied to real-world tasks in our signal processing community for speech recognition [13] and was followed by computer vision, natural language processing, robotics, speech synthesis, and image rendering [14]. Although deep learning and other machine learning approaches have shown impressive empirical success, many issues remain unsolved. In contrast to conventional linear modeling methods, deep learning methods are typically not interpretable. Although deep learning methodologies achieve recognition accuracy similar to or better than humans in many applications, they consume much more training data, power, and computing resources. Furthermore, despite statistically impressive results, individual accuracy results are often unreliable. Additionally, most of the current deep learning models lack reasoning and explanation capabilities, making them susceptible to catastrophic failures or attacks without the ability to anticipate and prevent them.

Fundamental as well as applied research is needed to overcome these challenges. Developing interpretable deep learning models could be a breakthrough in machine learning to create new algorithms and methods that can overcome the limitations of machine learning systems in not being able to explain actions, decisions, and prediction outcomes to human users while promising to perceive, learn, decide, and act independently. By understanding and trusting the system's output, users can predict future behavior and understand the system's outputs. Machine learning systems should be capable of creating models that explain how the world works when neural networks and symbolic systems are integrated. Their prediction and decision-making processes will be interpretable in symbolic and natural language by them as they discover the underlying causes or logical rules that govern them.

New algorithms for reinforcement learning and unsupervised deep learning could be a breakthrough in machine learning research, which use weak or no training signals paired with inputs to guide the learning process. By interacting with adversarial environments and with themselves, reinforcement learning algorithms can allow machines to learn. However, unsupervised learning has remained the most challenging problem for which no satisfactory algorithm has been developed. There has been a significant delay in developing unsupervised learning techniques compared to supervised and reinforcement deep learning techniques. Recent developments in unsupervised learning enable training prediction systems without labels by utilizing sequential output structures and advanced optimization methods.

6.8.2 Biomedical Imaging

In today's world, a variety of imaging technologies provide great insights into the body's anatomical and functional processes, including magnetic resonance imaging (MRI), computed tomography (CT), positron emission tomography (PET), optical coherence tomography (OCT), and ultrasound. There are still fundamental trade-offs between these aspects due to operational, financial, and physical constraints, even though such imaging technologies have improved significantly over time regarding resolution, signal-to-noise ratio (SNR), and acquisition speed. Because of noise, technology-related artifacts, poor resolution, and contrast, the acquired data can be largely unusable in raw form. Due to its complexity, it is also challenging for scientists and clinicians to interpret and analyze biomedical imaging data effectively and efficiently. Biomedical imaging researchers are developing new and exciting ways to resolve issues associated with the imaging of the human body, helping clinicians, radiologists, pathologists, and clinical researchers visualize, diagnose, and understand various diseases.

6.8.3 Natural Language Processing

Although natural language processing is a powerful tool, it still has limitations and issues: homonyms and contextual words, synonyms, sarcasm and irony, ambiguous situations, speech or text errors, slang and colloquialisms, languages specific to a particular domain, and languages with low resources [15].

A machine learning system requires a staggering amount of training data to work correctly. As a result, NLP models become more intelligent as they are trained on more data. Despite this, data (and human language!) are only increasing, as are machine learning techniques and algorithms tailored to a particular problem. More research and new methods will be needed to improve all these problems. NLP techniques, algorithms, and models can be developed using advanced techniques like artificial neural networks and deep learning. We will likely be able to come up with solutions to some of these challenges shortly as they grow and strengthen. Many of the limitations of NLP processing can be significantly eased with SaaS text analysis platforms like MonkeyLearn. In addition to automating customer service processes and collecting customer feedback, MonkeyLearn's no-code tools offer huge NLP benefits to streamline customer service processes.

6.8.4 Robotics

This section describes some open challenges when any robot is designed for specified applications [16]. These challenges are as per below:

1. **Developing a Motion Plan:** A robot must reach from one point to another without getting stuck anywhere along the way. Since the robot's surrounding environment is always dynamic, it is still an open research question. The robots must fetch this information and adapt to changing environments. Open

research problems include obtaining information about environmental changes and working spaces and adapting to them.

2. **Multiple Usage:** Suppose you design a robot for sorting different equipment. Now you want to teach the same robot for another task, such as the delivery of equipment. Then you are required to design new equations of motion and singularities, etc. It is a problem of being under-constrained and how to deal with it. It's still an open problem in designing robots for multiple uses.
3. **Simultaneous Location Mapping:** The human brain knows about body movement when they enter any environment and adjust according to it [18]. Therefore, the human brain is capable of creating surrounding maps and situations. However, it is challenging for robots to make this adjustment because it is designed for a specific environment. Therefore, creating simultaneous locations and mapping for the robot to adjust and adapt to any environmental changes is hard. That's why it is still an open problem and challenging to design simultaneous location mapping.
4. **Location Identification:** Many robots don't know how to deal with it when they lose track of their location. The method needs to design to deal with this situation. The technique depends on usage and specific application of the robot. It's likely if I created my robot, which can travel in different locations in the room, but what would happen if I put this robot on the staircase?
5. **Object Identification and Haptic Feedback:** It is not done 100 % yet. The robot manipulators with haptic feedback or even manipulated natural world objects with the help of object recognition are nowhere near tasks performed by the human hand. For example, lots of research has been published for picking up a stationary object. However, what happens if I designed my robots to grasp bananas from the bucket, but if I ask the same robot to fetch an orange, I need more time to get it? Also, in many cases in healthcare, the robot's performance could be better to the level of acceptance mark with objects which are not stable.
6. **Depth and Position Estimation:** Robots with vision can poke objects and see them move pretty easily. Moving objects are difficult to estimate if the robot doesn't know its distance from the object. It's very much an open problem.
7. **Real-Time Environment Understanding:** Suppose you are driving a car on the road and see your friend walking on the street path. Then you applied your intelligence to change the car's direction near the roadside by seeing traffic on the road and applied a break to stop the car at the roadside. This type of intelligence must be designed for robots to make them more efficient.

6.8.5 Wireless Communications

We live in a fully connected society thanks to wireless communications, which enable tetherless connectivity between people and the Internet. With the introduction of advanced transmission technologies such as multicarrier transmission, channel-adaptive transmission, and multiple antenna transmission and reception

(MIMO), mass-offering mobile broadband (MBB) access to the Internet has been the dominant theme of wireless communications for the past two decades. As the Internet of Things (IoT) and Industry 4.0 emerge, wireless communications will face new technical challenges. For example, multisensory virtual reality and UltraHD video increase spectral efficiency and explore extreme frequency bands.

Future wireless systems must simultaneously accommodate rapidly growing enhanced MBB services, mission-critical equipment, and IoT devices. A high degree of reliability, low latency, and energy efficiency are required for advanced IoT applications. In addition, multidimensional sensing and accurate localization will be essential for human-centric services in the future. The computing, communication, and control operations in Industry 4.0 must be fully integrated with artificial intelligence and machine learning. Costa and Yang identified various wireless communication challenges mentioned below [17]:

1. **Security and privacy**
2. **Utilization of spectrum**
3. **Development of communication infrastructure**
4. **Enhancement in energy efficiency**
5. **Integration of wireless information and power transfer**
6. **Development of wireless access techniques**
7. **Analysis of dynamic architecture and network function**
8. **Coding and modulation**
9. **Resources and interference management**

References

1. W. Zhao, Y. Zhang, N. Wang, Soft robotics: research, challenges, and prospects. J. Robot. Mechatron. **33**(1), 45–68
2. D. Trivedi, C.D. Rahn, W.M. Kier, I.D. Walker, Soft robotics: Biological inspiration, state of the art, and future research. Appl. Bionics Biomech. **5**(3), 99–117 (2008)
3. M. Manca, F. Paternò, C. Santoro, E. Zedda, C. Braschi, R. Franco, A. Sale, The impact of serious games with humanoid robots on mild cognitive impairment older adults. Int. J. Hum.-Comput. Stud. **145**, 102509 (2021)
4. V. Bonnet, J. Mirabel, D. Daney, F. Lamiraux, M. Gautier, O. Stasse, Practical whole-body elasto-geometric calibration of a humanoid robot: application to the TALOS robot. Robot. Auton. Syst. **164**, 104365 (2023)
5. C. Esterwood, L.P. Robert Jr, Three Strikes and you are out!: The impacts of multiple human–robot trust violations and repairs on robot trustworthiness. Comput. Hum. Behav. **142**, 107658 (2023)
6. R. Wen, A. Hanson, Z. Han, T. Williams, Fresh start: encouraging politeness in Wakeword-driven human-robot interaction, in *2023 ACM/IEEE International Conference on Human-Robot Interaction (HRI)* Stockholm, Sweden (2023)
7. M.G. Catalano, G. Grioli, E. Farnioli, A. Serio, C. Piazza, A. Bicchi, Adaptive synergies for the design and control of the Pisa/IIT SoftHand. Int. J. Robot. Res. **33**(5), 768–782 (2014)
8. G. Lentini, A. Settimi, D. Caporale, M. Garabini, G. Grioli, L. Pallottino, M.G. Catalano, Bicchi, A. Alter-ego: a mobile robot with a functionally anthropomorphic upper body designed for physical interaction. IEEE Robot. Autom. Mag. **26**(4), 94–107 (2019)

9. What is HMI? https://www.inductiveautomation.com/resources/article/what-is-hmi. Accessed Feb 2023
10. The 13 Most Popular AI Software Products in 2023 (2023). https://viso.ai/deep-learning/ai-software/. Accessed Feb. 2023
11. The 15 Best AI Tools to Know (2022). https://builtin.com/artificial-intelligence/ai-tools. Accessed Feb 2023
12. Y.C. Eldar, A.O. Hero III, L. Deng, J. Fessler, J. Kovacevic, H.V. Poor, S. Young, Challenges and open problems in signal processing: panel discussion summary from ICASSP 2017 [panel and forum]. IEEE Signal Process. Mag. **34**(6), 8–23 (2017)
13. G. Hinton, L. Deng, D. Yu, G.E. Dahl, A.R. Mohamed, N. Jaitly, A. Senior, V. Vanhoucke, P. Nguyen, T.N. Sainath, B. Kingsbury, Deep neural networks for acoustic modeling in speech recognition: the shared views of four research groups. IEEE Signal Process. Mag. **29**(6), 82–97 (2012)
14. I. Goodfellow, B. Yoshua, A. Courville, *Deep Learning* (MIT Press, Cambridge, 2016)
15. Major Challenges of Natural Language Processing (NLP) (2023). https://monkeylearn.com/blog/natural-language-processing-challenges/. Accessed Feb 2023
16. Open Problems in Robotics. https://scottlocklin.wordpress.com/2020/07/29/open-problems-in-robotics/. Last Accessed Feb 2023
17. D.B. Da Costa, H.C. Yang, Grand challenges in wireless communications. Front. Commun. Netw. **1**, 1 (2020)
18. J.M. Gomez-Quispe, G. Pérez-Zuñiga, D. Arce, F. Urbina, S. Gibaja, R. Paredes, F. Cuellar, Non linear control system for humanoid robot to perform body language movements. Sensors **23**(1), 552 (2023)
19. T. Cádrik, P. Takáč, J. Ondo, P. Sinčák, M. Mach, F. Jakab, F. Cavallo, M. Bonaccorsi, Cloud-based robots and intelligent space teleoperation tools, in *Robot Intelligence Technology and Applications*, vol. 4 (Springer, Berlin/Heidelberg, 2017), pp. 599–610
20. L. Fiorini, R. Esposito, M. Bonaccorsi et al., Enabling personalised medical support for chronic disease management through a hybrid robot-cloud approach. Auton. Robot. **41**, 1263–1276 (2017). https://doi.org/10.1007/s10514-016-9586-9
21. Y. Ma, Y. Zhang, J. Wan et al., Robot and cloud-assisted multi-modal healthcare system. Cluster Comput. **18**, 1295–1306 (2015). https://doi.org/10.1007/s10586-015-0453-9
22. A. Manzi, L. Fiorini, R. Limosani, P. Sincak, P. Dario, F. Cavallo, Use case evaluation of a cloud robotics teleoperation system (short paper), in *Proceedings of the 2016 5th IEEE International Conference on Cloud Networking (Cloudnet)*, Pisa, Italy, 3–5 October 2016, pp. 208–211

Index

R. Thanki, P. Joshi, *Advanced Technologies for Industrial Applications*,
https://doi.org/10.1007/978-3-031-33238-8

R

S

T

W

CPSIA information can be obtained
at www.ICGtesting.com
Printed in the USA
LVHW060802170723
752609LV00006B/503

9 783031 332371